de Gruyter Textbook

Pavel Drábek · Gabriela Holubová
Elements of Partial Differential Equations

Pavel Drábek · Gabriela Holubová

Elements of Partial Differential Equations

Walter de Gruyter
Berlin · New York

Pavel Drábek
Department of Mathematics
Faculty of Applied Sciences
University of West Bohemia
Univerzitní 22
306 14 Pilsen
Czech Republic
e-Mail: pdrabek@kma.zcu.cz

Gabriela Holubová
Department of Mathematics
Faculty of Applied Sciences
University of West Bohemia
Univerzitní 22
306 14 Pilsen
Czech Republic
e-Mail: gabriela@kma.zcu.cz

∞ Printed on acid-free paper which falls within the guidelines of the ANSI to ensure permanence and durability.

Library of Congress Cataloging-in-Publication Data

```
Drabek, P. (Pavel), 1953−
  Elements of partial differential equations / by Pavel Drábek and
  Gabriela Holubová.
    p. cm. − (De gruyter textbooks)
  Includes bibliographical references and index.
  ISBN 978-3-11-019124-0 (pbk. : alk. paper)
  1. Differential equations, Partial − Textbooks.   I. Holubová,
Gabriela.   II. Title.
  QA374.D66  2007
  515'.353−dc22
                                                       2006102208
```

Bibliographic information published by the Deutsche Nationalbibliothek

The Deutsche Nationalbibliothek lists this publication in the Deutsche Nationalbibliografie; detailed bibliographic data are available in the Internet at http://dnb.d-nb.de.

ISBN 978-3-11-019124-0

© Copyright 2007 by Walter de Gruyter GmbH & Co. KG, 10785 Berlin, Germany.
All rights reserved, including those of translation into foreign languages. No part of this book may be reproduced in any form or by any means, electronic or mechanical, including photocopy, recording, or any information storage and retrieval system, without permission in writing from the publisher.

Printed in Germany.
Coverdesign: +malsy, kommunikation und gestaltung, Willich.
Printing and binding: Druckhaus »Thomas Müntzer«, Bad Langensalza.

Preface

Nowadays, there are hundreds of books (textbooks as well as monographs) devoted to partial differential equations which represent one of the most powerful tools of mathematical modeling of real-world problems. These books contain an enormous amount of material. This is, on the one hand, an advantage since due to this fact we can solve plenty of complicated problems. However, on the other hand, the existence of such an extensive literature complicates the orientation in this subject for the beginners. It is difficult for them to distinguish an important fact from a marginal information, to decide what to study first and what to postpone for later time.

Our book is addressed not only to students who intend to specialize in mathematics during their further studies, but also to students of engineering, economy and applied sciences. To understand our book, only basic facts from calculus and linear ordinary differential equations of the first and second orders are needed. We try to present *the first introduction to PDEs* and that is why our text is on a very elementary level. Our aim is to enable the reader to understand what the partial differential equation is, where it comes from and why it must be solved. We also want the reader to understand the basic principles which are valid for particular types of PDEs, and to acquire some classical methods of their solving. We limit ourselves only to the fundamental types of equations and the basic methods. At present, there are many "software packages" (MATLAB, MAPLE, MATHEMATICA) that can be used for solving a lot of special types of partial differential equations and can be very helpful, but these tools are used only as a "black box" without deeper understanding of the general principles and methods.

The structure of the text is as follows. In Chapter 1, we present the basic conservation and constitutive laws. Chapter 2 is devoted to the classification of PDEs and to boundary and initial conditions. Chapter 3 deals with the first-order linear PDEs. Chapters 4–9 study the simplest possible forms of the wave, diffusion and Laplace (Poisson) equations and explain standard methods how to find their solutions. In Chapter 10, we point out the general principles of the above mentioned main three types of second-order linear PDEs. Chapters 11–13 are devoted to the Laplace (Poisson), diffusion and wave equations in higher dimensions. For the sake of brevity, we restrict ourselves to equations in three dimensions with respect to the space variable.

We would like to point out that *we work with most of the notions only intuitively and avoid intentionally some precise definitions and exact proofs*. We believe that this way of exposition makes the text more readable and the main ideas and methods used become thus more lucid. Hence, the reader who is not a mathematician, will be not disturbed by technical hypotheses, which in concrete models are usually assumed to be satisfied. On the other hand, we trust that a mathematician will fill up these gaps easily or find the answers in more specialized literature. In order to guide the reader, we put the most important equations, formulas and facts in boxes, so that he/she could better distinguish them from intermediate steps. Further, the text contains many solved examples and illustrating figures. At the end of each chapter there are several exercises.

More complicated and theoretical ones are accompanied by hints, the exercises based on calculations include the expected solution. We want to emphasize that there may exist many different forms of a solution and the reader can easily check the correctness of his/her results by substituting them in the equation.

There are many textbooks on PDEs which contain a lot of examples and exercises. Some of them appear in several books parallelly and so in these cases it is difficult to judge who was the original author. According to our best knowledge, we refer to the sources from which some of our examples are taken, and, moreover, at the end of each chapter, we list the sources from which we borrowed some of our exercises. On the other hand, many standard procedures and methods from our exposition have been also widely used by many other authors and we do not refer to all of them in order not to disturb the flow of ideas.

In any case, with the limited extent of this text, we do not claim our book to be comprehensive. It is a selection which is subjective, but which—in our opinion—covers the minimum which should be understood by everybody who wants to use or to study the theory of PDEs more deeply. We deal only with the classical methods which are a necessary starting point for further and more advanced studies. We draw from our experience that springs not only from our teaching activities but also from our scientific work. Within the scope of the Research Plan MSM 4977751301, we deal with problems more complicated but still similar to those studied in this book and investigate them (together with our colleagues) both from the analytical and numerical points of view.

Finally, we would like to thank our colleagues Petr Girg, Petr Nečesal, Pavel Krejčí, Alois Kufner, Herbert Leinfelder, Luboš Pick, Josef Polák and Robert Plato, editor of deGruyter, for careful reading of the manuscript and valuable comments. Our special thanks belong to Jiří Jarník for correcting our English.

Pilsen, November 2006   Pavel Drábek and Gabriela Holubová

Contents

Preface v

1 Mathematical Models, Conservation and Constitutive Laws 1
   1.1 Basic Notions 1
   1.2 Evolution Conservation Law 3
   1.3 Stationary Conservation Law 4
   1.4 Conservation Law in One Dimension 4
   1.5 Constitutive Laws 6
   1.6 Exercises 8

2 Classification, Types of Equations, Boundary and Initial Conditions . . . 9
   2.1 Basic Types of Equations, Boundary and Initial Conditions 9
   2.2 Classification of Linear Equations of the Second Order 14
   2.3 Exercises 17

3 Linear Partial Differential Equations of the First Order 21
   3.1 Convection and Transport Equation 21
   3.2 Equations with Constant Coefficients 22
   3.3 Equations with Non-Constant Coefficients 28
   3.4 Exercises 32

4 Wave Equation in One Spatial Variable—Cauchy Problem in $\mathbb{R}$ 37
   4.1 String Vibrations and Wave Equation in One Dimension 37
   4.2 Cauchy Problem on the Real Line 40
   4.3 Wave Equation with Sources 49
   4.4 Exercises 53

5 Diffusion Equation in One Spatial Variable—Cauchy Problem in $\mathbb{R}$ . . . 57
   5.1 Diffusion and Heat Equations in One Dimension 57
   5.2 Cauchy Problem on the Real Line 58
   5.3 Diffusion Equation with Sources 65
   5.4 Exercises 68

6 Laplace and Poisson Equations in Two Dimensions 71
   6.1 Steady States and Laplace and Poisson Equations 71
   6.2 Invariance of the Laplace Operator, Its Transformation into Polar Coordinates 73
   6.3 Solution of Laplace and Poisson Equations in $\mathbb{R}^2$ 74
   6.4 Exercises 76

7 Solutions of Initial Boundary Value Problems for Evolution Equations .. 78
- 7.1 Initial Boundary Value Problems on Half-Line 78
- 7.2 Initial Boundary Value Problem on Finite Interval, Fourier Method ... 84
- 7.3 Fourier Method for Nonhomogeneous Problems 99
- 7.4 Transformation to Simpler Problems 103
- 7.5 Exercises 104

8 Solutions of Boundary Value Problems for Stationary Equations 113
- 8.1 Laplace Equation on Rectangle 113
- 8.2 Laplace Equation on Disc 115
- 8.3 Poisson Formula 117
- 8.4 Exercises 119

9 Methods of Integral Transforms 123
- 9.1 Laplace Transform 123
- 9.2 Fourier Transform 128
- 9.3 Exercises 134

10 General Principles 139
- 10.1 Principle of Causality (Wave Equation) 139
- 10.2 Energy Conservation Law (Wave Equation) 141
- 10.3 Ill-Posed Problem (Diffusion Equation for Negative t) 144
- 10.4 Maximum Principle (Heat Equation) 145
- 10.5 Energy Method (Diffusion Equation) 147
- 10.6 Maximum Principle (Laplace Equation) 148
- 10.7 Consequences of Poisson Formula (Laplace Equation) 150
- 10.8 Comparison of Wave, Diffusion and Laplace Equations 152
- 10.9 Exercises 152

11 Laplace and Poisson equations in Higher Dimensions 157
- 11.1 Invariance of the Laplace Operator 157
- 11.2 Green's First Identity 159
- 11.3 Properties of Harmonic Functions 161
- 11.4 Green's Second Identity and Representation Formula 164
- 11.5 Boundary Value Problems and Green's Function 166
- 11.6 Dirichlet Problem on Half-Space and on Ball 168
- 11.7 Exercises 174

12 Diffusion Equation in Higher Dimensions 178
- 12.1 Heat Equation in Three Dimensions 178
- 12.2 Cauchy Problem in $\mathbb{R}^3$ 179
- 12.3 Diffusion on Bounded Domains, Fourier Method 182
- 12.4 Exercises 192

Contents

13 Wave Equation in Higher Dimensions 195
   13.1 Membrane Vibrations and Wave Equation in Two Dimensions 195
   13.2 Cauchy Problem in $\mathbb{R}^3$—Kirchhoff's Formula 196
   13.3 Cauchy problem in $\mathbb{R}^2$ 199
   13.4 Wave with sources in $\mathbb{R}^3$ 202
   13.5 Characteristics, Singularities, Energy and Principle of Causality 204
   13.6 Wave on Bounded Domains, Fourier Method 208
   13.7 Exercises 224

14 Appendix 228
   14.1 Sturm-Liouville problem 228
   14.2 Bessel Functions 229

Some Typical Problems Considered in This Book 235

Notation 237

Bibliography 239

Index 241

1 Mathematical Models, Conservation and Constitutive Laws

The beginning and development of the theory of partial differential equations were connected with physical sciences and with the effort to describe some physical processes and phenomena in the language of mathematics as precisely (and simply) as possible. With the invasion of new branches of science, this mathematical tool found its usage also outside physics. The width and complexity of problems studied gave rise to a new branch called *mathematical modeling*. The theory of partial differential equations was set apart as a separate scientific discipline. However, studying partial differential equations still stays closely connected with the description—modeling—of physical or other phenomena.

1.1 Basic Notions

In our text, the notion of a *mathematical model* is understood as a mathematical problem whose solution describes the behavior of the studied system. In general, a mathematical model is a simplified mathematical description of a real-world problem. In our case, we will deal with models described by partial differential equations, that is, differential equations with two or more independent variables.

Studying natural, technical, economical, biological, chemical and even social processes, we observe two main tendencies: the tendency to achieve a certain *balance* between causes and consequences, and the tendency to break this balance. Thus, as a starting point for the derivation of many mathematical models, we usually use some law or principle that expresses such a balance between the so called *state quantities* and *flow quantities* and their spatial and time changes.

Let us consider a medium (body, liquid, gas, solid substance, etc.) that fills a domain

$$\Omega \subset \mathbb{R}^N.$$

Here N denotes the spatial dimension. In real situations, usually, $N = 3$, in simplified models, $N = 2$ or $N = 1$. We denote by

$$u = u(\boldsymbol{x}, t), \quad \boldsymbol{x} \in \Omega, \, t \in [0, T) \subset [0, +\infty)$$

the state function (scalar, vector or tensor) of the substance considered at a point $\boldsymbol{x}$ and time t. In further considerations, we assume u to be a scalar function. The flow function (vector function, in general) of the same substance will be denoted by

$$\boldsymbol{\phi} = \boldsymbol{\phi}(\boldsymbol{x}, t), \quad \boldsymbol{x} \in \Omega, \, t \in [0, T) \subset [0, +\infty).$$

The density of sources is usually described by a scalar function

$$f = f(\boldsymbol{x}, t), \quad \boldsymbol{x} \in \Omega, \, t \in [0, T) \subset [0, +\infty).$$

Let $\Omega_B \subset \Omega$ be an arbitrary inner subdomain of Ω. The integral

$$U(\Omega_B, t) = \int_{\Omega_B} u(\boldsymbol{x}, t)\, \mathrm{d}\boldsymbol{x}$$

represents the total amount of the quantity considered in the *balance domain* Ω_B at time t. The integral

$$U(\Omega_B, t_1, t_2) = \int_{t_1}^{t_2}\int_{\Omega_B} u(\boldsymbol{x}, t)\, \mathrm{d}\boldsymbol{x}\, \mathrm{d}t$$

then represents the total amount of the quantity in Ω_B and in the time interval $[t_1, t_2] \subset [0, T]$. (The set $\Omega_B \times [t_1, t_2]$ is called the space-time balance domain.)

In particular, if the state function $u(\boldsymbol{x}, t)$ corresponds to the mass density $\varrho(\boldsymbol{x}, t)$, then the integral

$$m(\Omega_B, t) = \int_{\Omega_B} \varrho(\boldsymbol{x}, t)\, \mathrm{d}\boldsymbol{x}$$

represents the mass of the substance in the balance domain Ω_B at time t, and the value

$$m(\Omega_B, t_1, t_2) = \int_{t_1}^{t_2}\int_{\Omega_B} \varrho(\boldsymbol{x}, t)\, \mathrm{d}\boldsymbol{x}\, \mathrm{d}t$$

corresponds to the total mass of the substance contained in Ω_B during the time interval $[t_1, t_2]$.

If we denote by $\partial \Omega_B$ the boundary of the balance domain Ω_B, then the boundary integral (surface integral in $\mathbb{R}^3$, curve integral in $\mathbb{R}^2$)

$$\Phi(\partial \Omega_B, t) = \int_{\partial \Omega_B} \boldsymbol{\phi}(\boldsymbol{x}, t) \cdot \boldsymbol{n}(\boldsymbol{x})\, \mathrm{d}S$$

represents the amount of the quantity "flowing through" the boundary $\partial \Omega_B$ in the direction of the outer normal $\boldsymbol{n}$ at time t, and, similarly,

$$\Phi(\partial \Omega_B, t_1, t_2) = \int_{t_1}^{t_2}\int_{\partial \Omega_B} \boldsymbol{\phi}(\boldsymbol{x}, t) \cdot \boldsymbol{n}(\boldsymbol{x})\, \mathrm{d}S\, \mathrm{d}t$$

corresponds to the total amount of the quantity flowing through $\partial \Omega_B$ in the direction of the outer normal $\boldsymbol{n}$ during the time interval $[t_1, t_2]$.

As we have stated above, the distribution of sources is usually described by a function $f = f(\boldsymbol{x}, t)$ corresponding to the source density at a point $\boldsymbol{x}$ and time t. The integral

$$F(\Omega_B, t_1, t_2) = \int_{t_1}^{t_2}\int_{\Omega_B} f(\boldsymbol{x}, t)\, \mathrm{d}\boldsymbol{x}\, \mathrm{d}t$$

then represents the total source production in Ω_B during the time interval $[t_1, t_2]$.

1.2 Evolution Conservation Law

If the time evolution of the system has to be taken into account, we speak about an *evolution* process. To derive a balance principle for such a process, we choose an arbitrary balance domain $\Omega_B \subset \Omega$ and an arbitrary time interval $[t_1, t_2] \subset [0, \infty)$. For simplicity, we consider a scalar state function $u = u(\boldsymbol{x}, t)$, a vector flow function $\boldsymbol{\phi} = \boldsymbol{\phi}(\boldsymbol{x}, t)$, and a scalar source function $f = f(\boldsymbol{x}, t)$.

The basic balance law says that the change of the total amount of the quantity u contained in Ω_B between times t_1 and t_2 must be equal to the total amount flowing across the boundary $\partial \Omega_B$ from time t_1 to t_2, and to the increase (or decrease) of the quantity produced by sources (or sinks) inside Ω_B during the time interval $[t_1, t_2]$. In the language of mathematics, we write it as

(1.1)
$$\int_{\Omega_B} u(\boldsymbol{x}, t_2)\, d\boldsymbol{x} - \int_{\Omega_B} u(\boldsymbol{x}, t_1)\, d\boldsymbol{x}$$
$$= -\int_{t_1}^{t_2}\!\!\int_{\partial \Omega_B} \boldsymbol{\phi}(\boldsymbol{x}, t) \cdot \boldsymbol{n}(\boldsymbol{x})\, dS\, dt + \int_{t_1}^{t_2}\!\!\int_{\Omega_B} f(\boldsymbol{x}, t)\, d\boldsymbol{x}\, dt.$$

The minus sign in front of the first term on the right-hand side corresponds to the fact that the flux is understood as positive in the outward direction. Equation (1.1) represents the *evolution conservation law in its integral (global) form*.

If we assume u to have continuous partial derivative with respect to t, we can write the difference $u(\boldsymbol{x}, t_2) - u(\boldsymbol{x}, t_1)$ as $\int_{t_1}^{t_2} \frac{\partial}{\partial t} u(\boldsymbol{x}, t)\, dt$ and change the order of integration on the left-hand side of (1.1):

(1.2)
$$\int_{t_1}^{t_2}\!\!\int_{\Omega_B} \frac{\partial}{\partial t} u(\boldsymbol{x}, t)\, d\boldsymbol{x}\, dt$$
$$= -\int_{t_1}^{t_2}\!\!\int_{\partial \Omega_B} \boldsymbol{\phi}(\boldsymbol{x}, t) \cdot \boldsymbol{n}(\boldsymbol{x})\, dS\, dt + \int_{t_1}^{t_2}\!\!\int_{\Omega_B} f(\boldsymbol{x}, t)\, d\boldsymbol{x}\, dt.$$

Since the time interval $[t_1, t_2]$ has been chosen arbitrarily, we can come (under the assumption of continuity in the time variable of all functions involved) to the expression

(1.3)
$$\int_{\Omega_B} \frac{\partial}{\partial t} u(\boldsymbol{x}, t)\, d\boldsymbol{x} = -\int_{\partial \Omega_B} \boldsymbol{\phi}(\boldsymbol{x}, t) \cdot \boldsymbol{n}(\boldsymbol{x})\, dS + \int_{\Omega_B} f(\boldsymbol{x}, t)\, d\boldsymbol{x}.$$

Now, if we assume $\boldsymbol{\phi}$ to be continuously differentiable in the spatial variables, we can use the Divergence Theorem, according to which we can write

$$\int_{\partial \Omega_B} \boldsymbol{\phi}(\boldsymbol{x}, t) \cdot \boldsymbol{n}(\boldsymbol{x})\, dS = \int_{\Omega_B} \operatorname{div} \boldsymbol{\phi}(\boldsymbol{x}, t)\, d\boldsymbol{x}.$$

If we substitute this relation into (1.3) and assume the continuity of all functions in spatial variables, we come to

(1.4)
$$\frac{\partial}{\partial t} u(\boldsymbol{x}, t) + \operatorname{div} \boldsymbol{\phi}(\boldsymbol{x}, t) = f(\boldsymbol{x}, t),$$

since the balance domain has been chosen arbitrarily as well. Equation (1.4) is a local version of (1.1) and expresses the *conservation law in its differential (local) form*. It is a single equation for two unknown functions u and $\boldsymbol{\phi}$. The sources f are usually given, however, they can depend on $\boldsymbol{x}, t$ also via the quantity u, that is, we can have $f = f(\boldsymbol{x}, t, u(\boldsymbol{x}, t))$. Thus the conservation law alone is not sufficient for the construction of a mathematical model.

1.3 Stationary Conservation Law

Sometimes we are not interested in the time evolution of the system considered. We say that we study the *stationary state* or *stationary behavior* of the system. It means that we suppose all quantities to be time-independent (they have zero time derivatives). In such cases, we use simplified versions of conservation laws. In particular, the global form of the stationary conservation law is given by

(1.5)
$$\int_{\partial \Omega_B} \boldsymbol{\phi}(\boldsymbol{x}) \cdot \boldsymbol{n}(\boldsymbol{x}) \, dS = \int_{\Omega_B} f(\boldsymbol{x}) \, d\boldsymbol{x},$$

and its local version has the form

(1.6)
$$\operatorname{div} \boldsymbol{\phi}(\boldsymbol{x}) = f(\boldsymbol{x})$$

(under the assumptions of continuity of all functions involved and their spatial derivatives).

1.4 Conservation Law in One Dimension

In some situations, we can assume that all significant changes proceed only in one direction (for instance, in modeling the convection in a wide tube, when we are not interested in the situation near the tube walls; or, conversely, in modeling the behavior of a thin string or a thin bar with constant cross-section). In such cases, we can reduce our model to one spatial dimension. Since the corresponding one-dimensional basic conservation law differs in some minor points from the general one (e.g., ϕ is now a scalar function), we state it here explicitly.

Let us consider a tube with constant cross-section A, an arbitrary segment $a \leq x \leq b$, a time interval $[t_1, t_2]$, and a quantity with density u. The conservation law says, again, that the change of quantity in the spatial segment $[a, b]$ between times t_1 and t_2

Section 1.4 Conservation Law in One Dimension

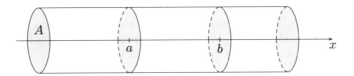

Figure 1.1 *An isolated tube with cross-section A; the quantities considered change only in the direction of the x-axis.*

equals the total flow at the point $x = a$ decreased by the total flow at the point $x = b$ from time t_1 to t_2, and to the total source balance in $[a, b]$ and $[t_1, t_2]$:

(1.7)
$$\int_a^b u(x, t_2)\, dx - \int_a^b u(x, t_1)\, dx$$
$$= \int_{t_1}^{t_2} (\phi(a, t) - \phi(b, t))\, dt + \int_{t_1}^{t_2} \int_a^b f(x, t)\, dx\, dt.$$

This equation represents the *one-dimensional conservation law in its integral (global) form* (cf. (1.1)). If the functions u and ϕ are smooth enough, we can proceed similarly to the general, multidimensional case, and obtain the differential formulation.

To be specific, if u has continuous partial derivative with respect to t and ϕ has continuous partial derivative with respect to x, equation (1.7) reduces to the form

$$\int_{t_1}^{t_2} \int_a^b [u_t(x, t) + \phi_x(x, t) - f(x, t)]\, dx\, dt = 0.$$

Since the intervals $[a, b]$ and $[t_1, t_2]$ have been chosen arbitrarily, the integrand must be identically equal to zero, thus

(1.8)
$$u_t(x, t) + \phi_x(x, t) = f(x, t).$$

Equation (1.8) is a local version of (1.7) and expresses the *one-dimensional conservation law in its differential (local) form* (cf. (1.4)).

If we model a one-dimensional stationary phenomenon, we use the stationary version of the previous conservation law, that is

(1.9)
$$\phi_x(x) = f(x),$$

which is actually an ordinary differential equation.

1.5 Constitutive Laws

In particular processes and phenomena, the state and flow functions (quantities) have their concrete terms and notation. For example, for the description of thermodynamic processes, we usually use some of the following parameters:

state parameters:	density	flow parameters:	velocity
	pressure		momentum
	temperature		tension
	specific internal energy		heat flux
	entropy		

The problem of mutual dependence or independence of these parameters is very complicated and it is connected with the choice of the mathematical model. Relations between the state quantity and the relevant flow quantity are usually based on the generalization of experimental observations and depend on the properties of the particular medium or material. They are usually called *constitutive laws* or *material relations*.

We close this chapter with several examples of fundamental physical processes and their mathematical models.

Example 1.1 (Convection and Transport.) The one-dimensional *convection model* describes the drift of a contaminant in a tube with a flowing liquid of constant speed. Here the state quantity is the concentration u of the contaminant, the flow quantity is its flux ϕ. Both these quantities are related by the linear constitutive law

$$\phi = cu$$

with a constant $c > 0$ corresponding to the velocity of the flowing liquid. Substituting this relation into the local conservation law, we obtain the *transport equation*

$$\boxed{u_t + cu_x = f.}$$

For details see Section 3.1.

□

Example 1.2 (Diffusion.) The *diffusion process* can be characterized by the state quantity u corresponding to the concentration of the diffusing substance and by its flux ϕ (the flow quantity). In this case, the constitutive law is so called Fick's law

$$\phi = -k \operatorname{grad} u,$$

which describes the fact that molecules tend to move from places of higher concentration to places with lower concentration. The material constant $k > 0$ is called the diffusion coefficient. This law together with the conservation law yields the *diffusion equation*

$$\boxed{u_t - k\Delta u = f.}$$

For detailed derivation see Sections 5.1 and 12.1.

□

Section 1.5 Constitutive Laws

Example 1.3 (Heat Flow.) In the model of the *heat flow*, the state quantity u corresponds to the temperature, and its multiple $c\rho u$ describes the density of the internal heat energy. Here c is the specific heat capacity and ρ the mass density (material constants) of the medium considered. The flow quantity is represented by the heat flux ϕ. The corresponding constitutive law is so called Fourier's law, which expresses the fact that the heat flux is directly proportional to the temperature gradient:

$$\phi = -K \text{ grad } u.$$

The constant $K > 0$ represents the heat (or thermal) conductivity (a material constant). Inserting the constitutive law directly into the conservation law (which is nothing else but the heat conservation law), we obtain the *heat equation*

$$\boxed{u_t - k\Delta u = f.}$$

(Here $k = \frac{K}{c\rho}$ is called the thermal diffusivity.) Notice that the heat and diffusion equations are identical! For detailed derivation see Sections 5.1 and 12.1. □

Example 1.4 (Wave motion.) The *wave motion* is described by the mass density of the medium and its displacement u (state quantities), and by its inner tension and momentum (flow quantities). The simplest model—the *wave equation*—has the form

$$\boxed{u_{tt} - c^2 \Delta u = f,}$$

where the constant c corresponds to the speed of the wave propagation. Here we use the conservation of mass and the conservation of momentum (Newton's law of motion), but we cannot use the standard scheme (to substitute the corresponding constitutive law into the conservation law). The derivation of the wave equation requires slightly different arguments and is treated in detail in Sections 4.1 and 13.1. □

Example 1.5 (Stationary processes.) The models of *stationary phenomena* can be derived separately (for example, as states with minimal energy), or we can understand them as special cases of the corresponding dynamical processes with time-independent quantities. Using the latter approach, the three previous equations are reduced to the *Poisson* (or *Laplace*, for $f = 0$) equation

$$\boxed{\Delta u = f.}$$

For details see Section 6.1. □

All the above situations are treated in detail in the subsequent chapters.

1.6 Exercises

1. How would you change the derivation of the conservation law (1.8) in the case of the tube with a variable cross-section $A = A(x)$?

2. Consider $u = u(x,t)$ to be the density of cars on a one-way road, and $\phi = \phi(x,t)$ the flux of the cars. Observe the following facts. If there is no car on the road, the flux is zero. There exists a critical value of the car density, say u_j, for which the jam occurs and the flux is zero again. And there must exist an optimal value u_m, $0 < u_m < u_j$, for which the flux is maximal. Try to formulate a simple constitutive law relating u and ϕ and to derive a basic (nonlinear) traffic model.

   [The traffic constitutive law can have the form $\phi = u(u_j - u)$ and the traffic model is then $u_t + u_j u_x - 2u u_x = 0$.]

3. Derive a nonlinear equation describing the behavior of bacteria in a one-dimensional tube under the assumption that the population growth obeys the logistic law $ru(1 - u/K)$. Here $u = u(x,t)$ denotes the concentration of the bacteria, r is a growth constant and K represents the carrying capacity. Use the basic conservation law with $\phi = -Du_x$ (the bacteria are diffusing inside the tube with the diffusion constant D) and $f = ru(1 - u/K)$ (the sources reflect the reproduction of the population).

   [The resulting model $u_t - Du_{xx} = ru(1 - u/K)$ is known as *Fisher's equation*.]

4. Derive the so called *Burgers equation*

$$u_t - Du_{xx} + uu_x = 0,$$

   which describes the coupling between nonlinear convection and diffusion in fluid mechanics. Use the one-dimensional conservation law (1.8) with no sources and the constitutive relation $\phi = -Du_x + \frac{1}{2}u^2$. Here, the first term represents the diffusion process, the latter term corresponds to the nonlinear transport (or convection).

5. Show that the Burgers equation $u_t - Du_{xx} + uu_x = 0$ can be transformed into the diffusion equation $\varphi_t - D\varphi_{xx} = 0$ using the *Cole-Hopf transform* $u = \psi_x$, $\psi = -2D \ln \varphi$.

6. Model an electric cable which is not well insulated from the ground, so that leakage occurs along its entire length. Consider that the voltage $V(x,t)$ and current $I(x,t)$ satisfy the relations $V_x = -LI_t - RI$ and $I_x = -CV_t - GV$, where L is the inductance, R is the resistance, C is the capacitance, and G is the leakage to the ground. Show that V and I both satisfy the *telegraph equation*

$$u_{xx} = LCu_{tt} + (RC + LG)u_t + RGu.$$

2 Classification, Types of Equations, Boundary and Initial Conditions

One of the main goals of the theory of partial differential equations is to express the unknown function of several independent variables from an identity where this function appears together with its partial derivatives. In the sequel, we keep the following notation: t denotes the time variable, $x, y, z, \ldots$ stand for the spatial variables. The *general partial differential equation* in 3D can be written as

$$F(x, y, z, t, u, u_x, u_y, u_z, u_t, u_{xy}, u_{xt}, \ldots) = 0,$$

where $(x, y, z) \in \Omega \subset \mathbb{R}^3$, $t \in I$, Ω is a given domain in $\mathbb{R}^3$ and $I \subset \mathbb{R}$ is a time interval. If $\boldsymbol{F}$ is a vector-valued function, $\boldsymbol{F} = (F_1, \ldots, F_m)$, and we look for several unknown functions $u = u(x, y, z, t)$, $v = v(x, y, z, t)$, $\ldots$, then

$$\boldsymbol{F}(x, y, z, t, u, u_x, \ldots, v, v_x, \ldots) = \boldsymbol{0}$$

is a *system of partial differential equations*. It is clear that these relations can be, in general, very complicated, and only some of their particular cases can be successfully studied by a mathematical theory. That is why it is important to know how to recognize these types and to distinguish them.

2.1 Basic Types of Equations, Boundary and Initial Conditions

Partial differential equations can be classified from various points of view. If time t is one of the independent variables of the searched-for function, we speak about *evolution* equations. If it is not the case (the equation contains only spatial independent variables), we speak about *stationary* equations. The highest order of the derivative of the unknown function in the equation determines the *order* of the equation. If the equation consists only of a linear combination of u and its derivatives (for example, it does not contain products as uu_x, $u_x u_{xy}$, etc.), we speak about a *linear* equation. Otherwise, it is a *nonlinear* equation. A linear equation can be written symbolically by means of a linear differential operator L, i.e., the operator with the property

$$L(\alpha u + \beta v) = \alpha L(u) + \beta L(v),$$

where α, β are real constants and u, v are real functions. The equation

$$L(u) = 0$$

is called *homogeneous*, the equation

$$L(u) = f,$$

where f is a given function, is called *nonhomogeneous*. The function f represents the "right-hand side" of the equation.

According to the above-mentioned aspects, we can classify the following equations:

1. The transport equation in one spatial variable:

$$u_t + u_x = 0$$

is evolution, of the first order, linear with $L(u) := u_t + u_x$, homogeneous.

2. The Laplace equation in three spatial variables:

$$\Delta u := u_{xx} + u_{yy} + u_{zz} = 0$$

is stationary, of the second order, linear with $L(u) := u_{xx} + u_{yy} + u_{zz}$, homogeneous.

3. The Poisson equation in two spatial variables:

$$\Delta u := u_{xx} + u_{yy} = f,$$

where $f = f(x,y)$ is a given function, is stationary, of the second order, linear with $L(u) := u_{xx} + u_{yy}$, nonhomogeneous.

4. The wave equation with interaction in one spatial variable:

$$u_{tt} - u_{xx} + u^3 = 0$$

is evolution, of the second order, nonlinear. The interaction is represented by the term u^3.

5. The diffusion equation in one spatial variable:

$$u_t - u_{xx} = f$$

is evolution, of the second order, linear with $L(u) := u_t - u_{xx}$, nonhomogeneous.

6. The equation of the vibrating beam:

$$u_{tt} + u_{xxxx} = 0$$

is evolution, of the fourth order, linear with $L(u) := u_{tt} + u_{xxxx}$, homogeneous.

7. The Schrödinger equation (a special case):

$$u_t - iu_{xx} = 0$$

is evolution, of the second order, linear with $L(u) := u_t - iu_{xx}$, homogeneous (here i is the imaginary unit: $i^2 = -1$).

Section 2.1 Basic Types of Equations, Boundary and Initial Conditions

8. The equation of a disperse wave:

$$u_t + u u_x + u_{xxx} = 0$$

is evolution, of the third order, nonlinear.

A function u is called a *solution* of a partial differential equation if, when substituted (together with its partial derivatives) into the equation, the latter becomes an identity. As in the case of ordinary differential equations, a solution of the partial differential equation is not determined uniquely. Thus we come to the notion of the *general solution*.

Let us notice the difference between the general solution of an ordinary differential equation (ODE) and the general solution of a partial differential equation (PDE): the general solution of an ODE includes *arbitrary constants* (their number is given by the order of the equation); the general solution of PDE includes *arbitrary functions*. This fact is illustrated by the examples below (cf. Strauss [19]).

Example 2.1 Let us search for a function of two variables $u = u(x, y)$ satisfying the equation

(2.1) $$u_{xx} = 0.$$

This problem can be solved by direct integration of equation (2.1). Since we integrate with respect to x, the integration "constant" can depend, in general, on y. From (2.1) it follows that

$$u_x(x, y) = f(y)$$

and, by further integration,

$$u(x, y) = f(y)x + g(y).$$

We thus obtain a general solution of equation (2.1), and f and g are arbitrary functions of the variable y. □

Example 2.2 Let us search for a function $u = u(x, y)$ satisfying the equation

(2.2) $$u_{yy} + u = 0.$$

Similarly to the case of the ODE for the unknown function $v = v(t)$,

$$v'' + v = 0,$$

when the general solution is a function $v(t) = A \cos t + B \sin t$ with arbitrary constants $A, B \in \mathbb{R}$ (the reader is asked to explain why), the general solution of equation (2.2) has the form

$$u(x, y) = f(x) \cos y + g(x) \sin y,$$

where f and g are arbitrary functions of the variable x. □

Example 2.3 Let us search for a function $u = u(x, y)$ satisfying the equation

(2.3)
$$u_{xy} = 0.$$

Integrating (2.3) with respect to y, we obtain

$$u_x = f(x)$$

(f is an arbitrary "constant" depending on x). Further integration with respect to x then leads to

$$u(x, y) = F(x) + G(y),$$

where $F' = f$. Functions F and G are again arbitrary. If we look for twice continuously differentiable function u, then its second partial derivatives are exchangeable. Hence we can integrate (2.3) first with respect to x and then with respect to y. Thus both F and G must be differentiable.

□

As in ODEs, a single PDE does not provide sufficient information to enable us to determine its solution uniquely. For the unique determination of a solution, we need further information. In the case of *stationary equations*, it is usually *boundary conditions* which, together with the equation, form a *boundary value problem*. For example,

$$\begin{cases} u_{xx} + u_{yy} = 0, & (x, y) \in B(\mathbf{0}, 1) = \{(x, y) \in \mathbb{R}^2 : x^2 + y^2 < 1\}, \\ u(x, y) = 0, & (x, y) \in \partial B(\mathbf{0}, 1) = \{(x, y) \in \mathbb{R}^2 : x^2 + y^2 = 1\} \end{cases}$$

forms a homogeneous Dirichlet boundary value problem for the Laplace equation. If, in general, Ω is a bounded domain in $\mathbb{R}^3$, we distinguish the following basic types of (linear) boundary conditions.

The *Dirichlet* boundary condition:

$$\boxed{u(x, y, z) = g(x, y, z), \quad (x, y, z) \in \partial\Omega,}$$

the *Neumann* boundary condition:

$$\boxed{\frac{\partial u}{\partial n}(x, y, z) = g(x, y, z), \quad (x, y, z) \in \partial\Omega,}$$

the *Robin* (sometimes called also *Newton*) boundary condition:

$$\boxed{A\frac{\partial u}{\partial n}(x, y, z) + Bu(x, y, z) = g(x, y, z), \quad (x, y, z) \in \partial\Omega,}$$

where $\frac{\partial u}{\partial n}$ denotes the derivative with respect to the outer normal n to the boundary (surface) of the domain Ω; A, B, $A^2 + B^2 \neq 0$, are given constants. If on various parts of the boundary $\partial\Omega$ different types of boundary conditions are given, we speak about

a problem with *mixed* boundary conditions. In the case $g \equiv 0$, the boundary conditions are called *homogeneous*, otherwise they are *nonhomogeneous*. In one dimension, that is in the case of problems on the interval $\Omega = (a, b)$, the boundary $\partial\Omega$ consists of two points $x = a$, $x = b$. Then, for example, the nonhomogeneous Neumann boundary conditions reduce to
$$-u_x(a) = g_1, \quad u_x(b) = g_2.$$
On an unbounded domain, for example, on the interval $\Omega = (0, \infty)$, where it is not possible to speak about a value of the given function at the point "infinity", the homogeneous Dirichlet boundary condition has the form
$$u(0) = 0, \quad \lim_{x \to \infty} u(x) = 0.$$

In the case of *evolution equations*, we usually deal, besides boundary conditions, also with *initial conditions* which, together with the equation and the boundary conditions, form an *initial boundary value problem*. For example,
$$\begin{cases} u_{tt} = u_{xx}, & t \in (0, \infty), \ x \in (0, 1), \\ u(0, t) = u(1, t) = 0, \\ u(x, 0) = \varphi(x), \ u_t(x, 0) = \psi(x), \end{cases}$$
forms an initial boundary value problem for the one-dimensional wave equation. Here the boundary conditions are the homogeneous Dirichlet ones. The function φ denotes the initial displacement and ψ stands for the initial velocity at a given point x. The derivative u_t at time $t = 0$ is understood as the derivative from the right. If we look for the so-called classical solution, the functions φ and ψ are supposed to be continuous and also the function u is continuous (in fact, even the partial derivatives of the second order are continuous). That is why the boundary and initial conditions must satisfy the *compatibility conditions*
$$\varphi(0) = \varphi(1) = 0.$$

By a *solution* (*classical solution*) of the initial boundary (or boundary) value problem, we understand a function which satisfies the equation as well as the boundary and initial conditions pointwise. In particular, the solution must be differentiable up to the order of the equation. These requirements can be too strong and thus the notion of a solution of a PDE (or of a system of PDEs) is often understood in another (generalized) sense. In our text, we confine ourselves mainly to searching for classical solutions. However, in several examples in later chapters we handle more general situations and we will notify the reader about that.

Another notion which we introduce in this part, is that of a *well-posed* boundary (or initial boundary) value problem. The problem is called *well-posed* if the following *three* conditions are satisfied:

(i) a solution of the problem *exists*;

(ii) the solution of the problem is determined *uniquely*;

(iii) the solution of the problem is *stable* with respect to the *given data*, which means that a "small change" of initial or boundary conditions, right-hand side, etc., causes only a "small change" of the solution.

The last condition concerns especially models of physical problems, since the given data can never be measured with absolute accuracy. However, the question left in the definition of stability is what does "very small" or "small" change mean. The answer depends on the particular problem and, at this moment, we put up with only an intuitive understanding of this notion.

The contrary of a well-posed problem is the *ill-posed* problem, i.e., a problem which does not satisfy at least one of the three previous requirements. If the solution exists but the uniqueness is not ensured, the problem can be underdetermined. Conversely, if the solution does not exist, it can be an overdetermined problem. An underdetermined problem, overdetermined problem, as well as unstable problem can, however, make real sense. Further, it is worth mentioning that the notion of a well-posed problem is closely connected to the definition of a solution. As we will see later, the wave equation with non-smooth initial conditions is, in the sense of the classical solution defined above, an ill-posed problem, since its classical solution does not exist. However, if we consider the solution in a more general sense, the problem becomes well-posed, the generalized solution exists, it is unique and stable with respect to "small" changes of given data.

2.2 Classification of Linear Equations of the Second Order

In this section we state the classification of the basic types of PDEs of the second order that can be found most often in practical models.

The basic types of linear evolution equations of the second order are the *wave equation* (in one spatial variable):

$$u_{tt} - u_{xx} = 0 \quad (c = 1),$$

which is of *hyperbolic type*, and the *diffusion equation* (in one spatial variable):

$$u_t - u_{xx} = 0 \quad (k = 1),$$

which is of *parabolic type*. The basic type of the linear stationary equation of the second order (in two spatial variables) is the *Laplace equation*:

$$u_{xx} + u_{yy} = 0,$$

which is of *elliptic type*. Formal analogues of these PDEs are equations of conics in the plane: the equation of a hyperbola, $t^2 - x^2 = 1$, the equation of a parabola, $t - x^2 = 1$, and the equation of an ellipse (here we mention its special case—a circle), $x^2 + y^2 = 1$.

Let us consider a general linear homogeneous PDE of the second order

(2.4) $$a_{11}u_{xx} + 2a_{12}u_{xy} + a_{22}u_{yy} + a_1 u_x + a_2 u_y + a_0 u = 0$$

Section 2.2 Classification of Linear Equations of the Second Order

with two independent variables x, y and with six real coefficients that can depend on x and y. Let us denote by

$$A = \begin{bmatrix} a_{11} & a_{12} \\ a_{12} & a_{22} \end{bmatrix}$$

the matrix formed by the coefficients at the partial derivatives of the second order. It is possible to show that there exists a linear transformation of variables x, y, which reduces equation (2.4) to one of the following forms. In this respect, an important role is played by the determinant $\det A$ of the matrix A.

(i) *Elliptic form*: If $\det A > 0$, that is $a_{11}a_{22} > a_{12}^2$, the equation is reducible to the form

$$u_{xx} + u_{yy} + \cdots = 0,$$

where the dots represent terms with derivatives of lower orders.

(ii) *Hyperbolic form*: If $\det A < 0$, that is $a_{11}a_{22} < a_{12}^2$, the equation is reducible to the form

$$u_{xx} - u_{yy} + \cdots = 0.$$

The dots stand for the terms with derivatives of lower orders.

(iii) *Parabolic form*: If $\det A = 0$, that is $a_{11}a_{22} = a_{12}^2$, the equation is reducible to the form

$$u_{xx} + \cdots = 0, \quad (\text{or } u_{yy} + \cdots = 0),$$

unless $a_{11} = a_{12} = a_{22} = 0$. Here, again, the dots represent terms with derivatives of lower orders.

Finding the corresponding transformation relations and reducing the equation is based on the same idea as the analysis of conics in analytic geometry. For simplicity, let us consider only the principal terms in the equation, that is, let $a_1 = a_2 = a_0 = 0$, and let us normalize the equation by $a_{11} = 1$. If, moreover, we denote $\partial_x = \partial/\partial x$, $\partial_y^2 = \partial^2/\partial y^2$, etc., we can write equation (2.4) as

$$(\partial_x^2 + 2a_{12}\partial_x\partial_y + a_{22}\partial_y^2)u = 0$$

and, by formally completing the square, we convert it to

(2.5) $$(\partial_x + a_{12}\partial_y)^2 u + (a_{22} - a_{12}^2)\partial_y^2 u = 0.$$

Further, let us consider the elliptic case $a_{22} > a_{12}^2$, and denote $b = (a_{22} - a_{12}^2)^{1/2}$, which means $b \in \mathbb{R}$. We introduce new independent variables ξ and η by

$$x = \xi, \quad y = a_{12}\xi + b\eta.$$

The transformed derivatives assume the form

$$\partial_\xi = \partial_x + a_{12}\partial_y, \quad \partial_\eta = b\partial_y$$

(you can prove it using the chain rule), and equation (2.5) becomes

$$\partial_\xi^2 u + \partial_\eta^2 u = 0$$

or, equivalently,
$$u_{\xi\xi} + u_{\eta\eta} = 0.$$
In the remaining two cases, we would proceed analogously (in the parabolic case we have $b = 0$, and in the hyperbolic case, $b = i(a_{12}^2 - a_{22})^{1/2} \in \mathbb{C}$).

Example 2.4 We determine types of the following equations:

(a) $\quad u_{xx} - 3u_{xy} = 0,$

(b) $\quad 3u_{xx} - 6u_{xy} + 3u_{yy} + u_x = 0,$

(c) $\quad 2u_{xx} + 2u_{xy} + 3u_{yy} = 0.$

In terms of the previous explanation, we decide according to the sign of $\det \boldsymbol{A} = a_{11}a_{22} - a_{12}^2$. Thus, in case (a), we obtain $\det \boldsymbol{A} = -9/4 < 0$ and the equation is of hyperbolic type. In case (b), we have $\det \boldsymbol{A} = 0$, and thus the equation is of parabolic type. In case (c), we have $\det \boldsymbol{A} = 5 > 0$, and the equation is of elliptic type.

$\square$

If $\boldsymbol{A}$ is a function of x and y (i.e., the equation has nonconstant coefficients), then the type of the equation may be different in different parts of the xy-plane. See the following two examples.

Example 2.5 We find regions of the xy-plane where the equation
$$xu_{xx} - u_{xy} + yu_{yy} = 0$$
is of elliptic, hyperbolic, or parabolic type, respectively.

In this case the coefficients depend on x and y and we obtain $\det \boldsymbol{A} = a_{11}a_{22} - a_{12}^2 = xy - \frac{1}{4}$. The equation is thus of parabolic type on the hyperbola $xy = \frac{1}{4}$, of elliptic type in two convex regions $xy > \frac{1}{4}$, and of hyperbolic type in the connected region $xy < \frac{1}{4}$. The reader is invited to sketch a picture of corresponding regions.

$\square$

Example 2.6 Again, we find regions of the xy-plane where the equation
$$-x^2 u_{xx} + 2xy u_{xy} + (1+y)u_{yy} = 0$$
is of elliptic, hyperbolic, or parabolic type, respectively.

This time we have $\det \boldsymbol{A} = a_{11}a_{22} - a_{12}^2 = -x^2(1+y) - x^2 y^2$. The equation is thus of hyperbolic type in the whole plane except the axis y, where it is of parabolic type.

$\square$

Remark 2.7 In a similar way we can classify linear PDEs of the second order with an arbitrary finite number of variables N. The coefficient matrix $\boldsymbol{A}$ is then of type $N \times N$. The type of the equation is related to definiteness of the matrix $\boldsymbol{A}$ and can be determined by signs of its eigenvalues:

(i) the equation is of *elliptic* type, if the eigenvalues of A are all positive or all negative (i.e., A is positive or negative definite);

(ii) the equation is of *parabolic* type, if A has exactly one zero eigenvalue and all the other eigenvalues have the same sign (i.e., A is a special case of a positive or negative semidefinite matrix);

(iii) the equation is of *hyperbolic* type, if A has only one negative eigenvalue and all the others are positive, or A has only one positive eigenvalue and all the others are negative (i.e., A is a special case of an indefinite matrix);

(iv) the equation is of *ultrahyperbolic* type, if A has more than one positive eigenvalue and more than one negative eigenvalue, and no zero eigenvalues (i.e., A is indefinite).

Notice that the matrix A is symmetric, since we consider exchangeable second partial derivatives, and thus all its eigenvalues have to be real.

2.3 Exercises

1. Determine which of the following operators are linear.

    (a) $u \mapsto y\, u_x + u_y$,
    (b) $u \mapsto u u_x + u_y$,
    (c) $u \mapsto u_x^3 + u_y$,
    (d) $u \mapsto u_x + u_y + x + y$,
    (e) $u \mapsto (x^2 + y^2)(\sin y)\, u_x + x^3\, u_{yxy} + [\arccos(xy)] u$.

    [a,d,e]

2. In the following equations, determine their order and whether they are nonlinear, linear nonhomogeneous, or linear homogeneous. Explain your reasoning.

    (a) $u_t - 3 u_{xx} + 5 = 0$,
    (b) $u_t - u_{xx} + xt^3 u = 0$,
    (c) $u_t + u_{xxt} + u^2 u_x = 0$,
    (d) $u_{tt} - 4 u_{xx} + x^4 = 0$,
    (e) $i u_t - u_{xx} + x^3 = 0$,
    (f) $u_x(1 + u_x^2)^{-1/2} + u_y(1 + u_y^2)^{-1/2} = 0$,
    (g) $e^x u_x + u_y = 0$,
    (h) $u_t + u_{xxxx} + \sqrt[3]{1+u} = 0$.
    (i) $u_{xx} + e^t u_{tt} = u \cos x$.

    [linear: a,b,d,e,g,i]

3. Verify that the function $u(x,y,z) = f(x)g(y)h(z)$ solves the equation $u^2 u_{xyz} = u_x u_y u_z$ for arbitrary (differentiable) functions f, g and h of a single real variable.

4. Show that the nonlinear equation $u_t = u_x^2 + u_{xx}$ can be transformed to the diffusion equation $u_t = u_{xx}$ by using the transformation of the dependent variable $w = e^u$ (i.e., introducing a new unknown function w).

5. What are the types of the following equations?

    (a) $u_{xx} - u_{xy} + 2u_y + u_{yy} - 3u_{yx} + 4u = 0$,

    [hyperbolic]

    (b) $9u_{xx} + 6u_{xy} + u_{yy} + u_x = 0$.

    [parabolic]

6. Classify the equation
$$u_{xx} + 2ku_{xt} + k^2 u_{tt} = 0, \quad k \neq 0.$$
Use the transformation $\xi = x + bt$, $\tau = x + dt$ of the independent variables with unknown coefficients b and d such that the equation is reduced to the form $u_{\xi\xi} = 0$. Find a solution of the original equation.

[equation of parabolic type; $u(x,t) = f(x - \tfrac{t}{k})x + g(x - \tfrac{t}{k})$]

7. Classify the equation
$$xu_{xx} - 4u_{xt} = 0$$
in the domain $x > 0$. Solve this equation by a *nonlinear* substitution $\tau = t$, $\xi = t + 4\ln x$.

[equation of hyperbolic type; $u = e^{-t/4} f(t + 4\ln x) + g(t)$]

8. Show that the equation
$$u_{tt} - c^2 u_{xx} + au_t + bu_x + du = f(x,t)$$
can be transformed to the form
$$w_{\xi\tau} + kw = g(\xi,\tau), \quad w = w(\xi,\tau)$$
by substitutions $\xi = x - ct$, $\tau = x + ct$ and $u = we^{\alpha\xi + \beta\tau}$ for a suitable choice of constants α and β.

[$\alpha = \tfrac{b+ac}{4c^2}, \beta = \tfrac{b-ac}{4c^2}$]

9. Classify the equation
$$u_{xx} - 6u_{xy} + 12u_{yy} = 0.$$
Find a transformation of independent variables which converts it into the Laplace equation.

[equation of elliptic type; $\xi = x, \eta = \sqrt{3}x + \tfrac{1}{\sqrt{3}}y$]

Section 2.3 Exercises

10. Determine in which regions of the xy-plane the following equations are elliptic, hyperbolic, or parabolic.

    (a) $2u_{xx} + 4u_{yy} + 4u_{xy} - u = 0$,

    (b) $u_{xx} + 2y\, u_{xy} + u_{yy} + u = 0$,

    (c) $\sin(xy)u_{xx} - 6u_{xy} + u_{yy} + u_y = 0$,

    (d) $u_{xx} - \cos(x)u_{xy} + u_{yy} + u_y - u_x + 5u = 0$,

    (e) $(1+x^2)u_{xx} + (1+y^2)u_{yy} + xu_x + yu_y = 0$,

    (f) $e^{2x}u_{xx} + 2e^{x+y}u_{xy} + e^{2y}u_{yy} + (e^{2y} - e^{x+y})u_y = 0$,

    (g) $u_{xx} - 2\sin x\, u_{xy} - \cos^2 x\, u_{yy} - \cos x\, u_y = 0$,

    (h) $e^{xy}u_{xx} + (\cosh x)u_{yy} + u_x - u = 0$,

    (i) $[\log(1 + x^2 + y^2)]u_{xx} - [2 + \cos x]u_{yy} = 0$.

11. Try to find PDEs whose general solutions are of the form

    (a) $u(x,y) = \varphi(x+y) + \psi(x-2y)$,

    (b) $u(x,y) = x\varphi(x+y) + y\psi(x+y)$,

    (c) $u(x,y) = \frac{1}{x}(\varphi(x-y) + \psi(x+y))$.

    Here φ, ψ are arbitrary differentiable functions.

    [(a) $2u_{xx} - u_{xy} - u_{yy} = 0$, (b) $u_{xx} - 2u_{xy} + u_{yy} = 0$, (c) $x(u_{xx} - u_{yy}) + 2u_x = 0$.]

12. Consider the *Tricomi equation*

    $$yu_{xx} + u_{yy} = 0.$$

    Show that this equation is

    (a) elliptic for $y > 0$ and can be reduced to

    $$u_{\xi\xi} + u_{\eta\eta} + \frac{1}{3\eta}u_\eta = 0$$

    using the transformation $\xi = x$, $\eta = \frac{2}{3}y^{3/2}$;

    (b) hyperbolic for $y < 0$ and can be reduced to

    $$u_{\xi\eta} - \frac{1}{6(\xi - \eta)}(u_\xi - u_\eta) = 0$$

    using the transformation $\xi = x - \frac{2}{3}(-y)^{3/2}$, $\eta = x + \frac{2}{3}(-y)^{3/2}$.

13. Show that the equation

$$u_{xx} + yu_{yy} + \frac{1}{2}u_y = 0$$

can be reduced to the (canonical) form $u_{\xi\eta} = 0$ in the region where it is of hyperbolic type. Use this result to show that in the hyperbolic region it has the general solution

$$u(x, y) = f(x + 2\sqrt{-y}) + g(x - 2\sqrt{-y}),$$

where f and g are arbitrary functions.

14. Determine whether the following three-dimensional equations are of elliptic, hyperbolic, or parabolic type (determine the eigenvalues of the corresponding matrix A—see Remark 2.7):

    (a) $u_{xx} + 2u_{yz} + (\cos x)u_z - e^{y^2}u = \cosh z$,
    (b) $u_{xx} + 2u_{xy} + u_{yy} + 2u_{zz} - (1 + xy)u = 0$,
    (c) $7u_{xx} - 10u_{xy} - 22u_{yz} + 7u_{yy} - 16u_{xz} - 5u_{zz} = 0$,
    (d) $e^z u_{xy} - u_{xx} = \log(x^2 + y^2 + z^2 + 1)$.

15. Determine the regions of the xyz-space where

$$u_{xx} - 2x^2 u_{xz} + u_{yy} + u_{zz} = 0$$

is of hyperbolic, elliptic, or parabolic type.

Exercise 5 is taken from Strauss [19], 6–9 from Logan [14], 11, 12 from Stavroulakis and Tersian [18], and 13–15 from Zauderer [22].

3 Linear Partial Differential Equations of the First Order

3.1 Convection and Transport Equation

Transport Equation. We start with the derivation of the so called *transport equation*. We come from the basic one-dimensional conservation law (1.8) without sources (i.e., $f \equiv 0$), that is

$$u_t(x,t) + \phi_x(x,t) = 0,$$

and use the constitutive law according to which the flux density ϕ is proportional to the quantity u:

$$\phi = cu,$$

where c is a positive constant. After the substitution, we obtain the required *transport equation*

(3.1) $$\boxed{u_t + cu_x = 0.}$$

Such a model describes, for instance, the drift of a substance in a tube with a flowing liquid. Here the quantity u represents the concentration of the drift substance (say a contaminant) and the parameter c corresponds to the velocity of the flowing liquid. This model (often called the *convection model*) does not consider diffusion.

As we will show later, the solution of (3.1) is a function

(3.2) $$\boxed{u(x,t) = F(x - ct),}$$

where F is an arbitrary differentiable function. Such a solution is called the *right traveling wave*, since the graph of the function $F(x - ct)$ at a given time t is the graph of the function $F(x)$ shifted to the right by the value ct. Thus, with growing time, the profile $F(x)$ is moving without changes to the right at the speed c (see Figure 3.1).

If, in general, the velocity parameter c is negative, which means that the liquid and so the drifted substance flow to the *left* at speed $|c|$, the solution $u(x,t) = F(x - ct)$ is called *left traveling wave*.

If the flux density ϕ is a nonlinear function of the quantity u, then the conservation law (in case of $f \equiv 0$) has the form

(3.3) $$u_t + (\phi(u))_x = u_t + \phi'(u)u_x = 0.$$

This relation models nonlinear transport, which is, from the point of further analysis, more complicated.

Transport with Decay. The particle decay (radioactive decay of nuclei) can be described by the decay equation

$$\frac{du}{dt} = -\lambda u,$$

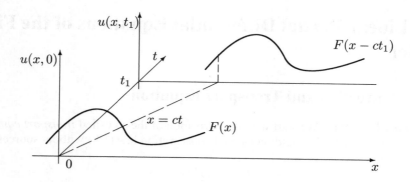

Figure 3.1 *Traveling wave.*

where u is the number of the so far nondecayed particles (nuclei) at time t and λ is a decay constant. The behavior of the substance (radioactive chemical) drift in a tube at speed c can be modeled by the equation

(3.4) $$u_t + cu_x = -\lambda u.$$

In this case we have again $\phi = cu$ and $f = -\lambda u$ represents the sources due to the decay.

In the next sections, we present several ways how to find the solutions of the above mentioned transport equations.

3.2 Equations with Constant Coefficients

Let us consider the general linear PDE of the first order with two independent variables and with *constant coefficients*

(3.5) $$au_x + bu_y = 0.$$

Here $u = u(x, y)$ is the unknown function, a, b are constants such that $a^2 + b^2 > 0$ (they are not both equal to zero). Solving equation (3.5) can be approached from various points of view. Below we present three basic methods.

Geometric Interpretation—Method of Characteristics. Let us denote $v = (a, b)$, $\nabla u = \operatorname{grad} u = (u_x, u_y)$. The left-hand side of equation (3.5) can be then considered as a scalar product

$$au_x + bu_y = v \cdot \nabla u,$$

Section 3.2 Equations with Constant Coefficients

and equation (3.5) can be interpreted in the following way: "the derivative of the function u in the direction of the vector v is equal to zero," or "the value of the function u does not change (is constant) in the direction of the vector v." In other words, u is constant on every line the directional vector of which is v (warning: this constant differs, in general, on different lines!).

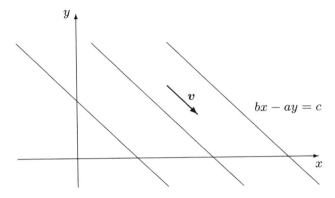

Figure 3.2 *Characteristic lines.*

Thus,
$$u(x,y) = f(bx - ay),$$
since the function $u(x,y)$ assumes the value $f(c)$ (and it is thus constant) on the given line $bx - ay = c$. Here f is an arbitrary differentiable real function. Lines described by $bx - ay = c$, $c \in \mathbb{R}$, are called the *characteristic lines*, or the *characteristics* of equation (3.5).

Example 3.1 We solve the equation

(3.6) $$2u_x - 3u_y = 0$$

with the initial condition

(3.7) $$u(0,y) = y^2.$$

On the basis of the previous text, we have
$$u(x,y) = f(-3x - 2y).$$

If we use the condition
$$y^2 = u(0,y) = f(-2y)$$

and substitute $w = -2y$, we obtain

$$f(w) = \frac{w^2}{4}.$$

Hence,

$$u(x, y) = \frac{(3x + 2y)^2}{4}$$

(see Figure 3.3).

Finally, we should verify that our function u is indeed a solution to the equation. Since

$$u_x = \frac{3}{2}(3x + 2y), \quad u_y = 3x + 2y,$$

after substituting them into the equation we find out that the left-hand side is equal to the right-hand side:

$$2u_x - 3u_y = 3(3x + 2y) - 3(3x + 4y) = 0,$$

and also $u(0, y) = y^2$.

$\square$

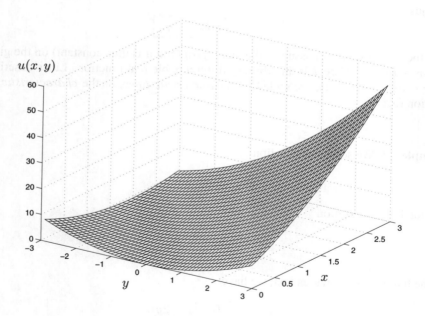

Figure 3.3 *The solution from Example 3.1.*

Section 3.2 Equations with Constant Coefficients

Remark 3.2 The solution of equation (3.6) satisfying the initial condition (3.7) is unique. Indeed, let $u_1 = u_1(x,y)$ and $u_2 = u_2(x,y)$ be two solutions of (3.6), (3.7). Then $w = u_1 - u_2$ solves (3.6) with the initial condition $w(0,y) = 0$. Hence $w(x,y) = f(-3x - 2y)$ and $0 = w(0,y) = f(-2y)$. This implies $f \equiv 0$, i.e., $w(x,y) \equiv 0$. So, u_1 and u_2 coincide.

Figure 3.4 depicts the function $u(x,t) = e^{-(3t+x)^2}$ which solves the equation

$$(3.8) \qquad u_t - 3u_x = 0$$

with the condition $u(x,0) = e^{-x^2}$. Similarly, Figure 3.5 shows the graph of the function $u(x,t) = -\sin(3t - x)$ which solves the equation

$$(3.9) \qquad u_t + 3u_x = 0$$

with the condition $u(x,0) = \sin x$. The reader is invited to find solutions of both (3.8), (3.9), and to notice how the initial conditions propagate along the characteristics.

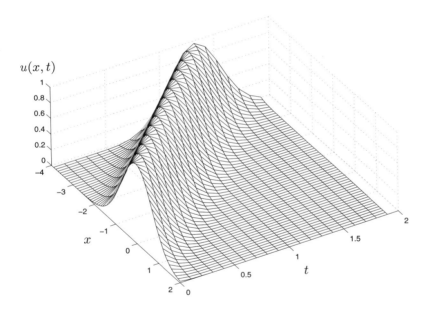

Figure 3.4 *Solution of* $u_t - 3u_x = 0$, $u(x,0) = e^{-x^2}$.

Coordinate Method. Again, let us suppose $a^2 + b^2 > 0$. We introduce a new rectangular coordinate system by the substitution

$$x' = ax + by, \quad y' = bx - ay.$$

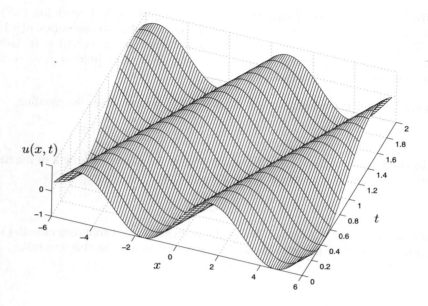

Figure 3.5 *Solution of* $u_t + 3u_x = 0$, $u(x,0) = \sin x$.

According to the chain rule for the derivative of a composite function, we have

$$u_x = \frac{\partial u}{\partial x} = \frac{\partial u}{\partial x'}\frac{\partial x'}{\partial x} + \frac{\partial u}{\partial y'}\frac{\partial y'}{\partial x} = au_{x'} + bu_{y'},$$

$$u_y = \frac{\partial u}{\partial y} = \frac{\partial u}{\partial x'}\frac{\partial x'}{\partial y} + \frac{\partial u}{\partial y'}\frac{\partial y'}{\partial y} = bu_{x'} - au_{y'}.$$

Equation (3.5), that is

$$au_x + bu_y = 0,$$

can be then written in the form

$$a^2 u_{x'} + abu_{y'} + b^2 u_{x'} - abu_{y'} = 0,$$

i.e.,

$$\underbrace{(a^2 + b^2)}_{\neq 0} u_{x'} = 0,$$

whence

$$u_{x'} = 0.$$

Thus, it follows that

$$u(x', y') = f(y'), \quad \text{i.e.,} \quad u(x,y) = f(bx - ay).$$

Section 3.2 Equations with Constant Coefficients

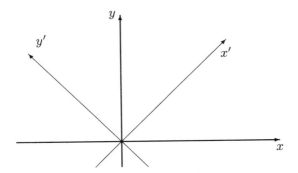

Figure 3.6 *Transformation of the coordinate system.*

We invite the reader to apply the coordinate method to find solutions of equations (3.8) and (3.9).

Method of Characteristic Coordinates. Let us consider the general form of the transport equation with decay

(3.10) $$u_t + cu_x + \lambda u = f(x,t),$$

where λ, c are constants and f is a given function representing external sources (the decay is characterized by the term λu—see Section 3.1). Since the transport equation describes the signal propagation at speed c, it is reasonable to try to transform the equation into a new, moving, coordinate system. We introduce thus new independent variables ξ and τ, called *characteristic coordinates*, by

$$\xi = x - ct, \quad \tau = t.$$

The variable ξ can be understood as a coordinate which propagates together with the signal. Using the chain rule, we derive

$$u_t = u_\xi \xi_t + u_\tau \tau_t = -cu_\xi + u_\tau,$$
$$u_x = u_\xi \xi_x + u_\tau \tau_x = u_\xi.$$

After substituting it into the original relation, we obtain the equation

(3.11) $$u_\tau + \lambda u = f(\xi + c\tau, \tau),$$

which can be further solved by methods of the theory of linear first-order ODEs. We illustrate this approach by the next example.

Example 3.3 Let us find the general solution of the transport equation

$$u_t + u_x - u = t.$$

We introduce the characteristic coordinates by the relations
$$\xi = x - t, \quad \tau = t$$
and, according to the instructions above, we transform the equation to the form
$$u_\tau - u = \tau.$$
Now, we treat this equation as an ODE with the variable τ and the parameter ξ. The solution of the corresponding homogeneous equation assumes the form
$$u_H(\xi, \tau) = g(\xi) e^\tau,$$
while a particular solution of the nonhomogeneous equation can be written as
$$u_P(\xi, \tau) = -(1 + \tau).$$
The general solution is then the sum $u = u_H + u_P$. Finally, we return to the original variables x and t:
$$u(x, t) = -(1 + t) + g(x - t) e^t.$$
The reader is asked to carry out all the above steps in detail and to check the correctness of the solution.

□

3.3 Equations with Non-Constant Coefficients

The method of characteristics, based on the geometric interpretation, can be used also in the case of equations with non-constant coefficients $a(x, y)$ and $b(x, y)$. The difference consists in the fact that the characteristics are no more straight lines but general curves. Instead of dealing with the general case, we prefer to illustrate the method by several examples.

Example 3.4 (Strauss [19]) We solve the equation

(3.12)
$$u_x + y u_y = 0.$$

Here we have $a = 1$, $b = y$, $v = (1, y)$, that is, the second component of the vector v depends on the variable y. The set of all vectors v in the xy plane can be depicted as in Figure 3.7.

The characteristics are thus the curves with the property that their tangents at a given point have the slope
$$\frac{dy}{dx} = y.$$
By solving this ODE we obtain the characteristics
$$y = ce^x,$$

Section 3.3 Equations with Non-Constant Coefficients

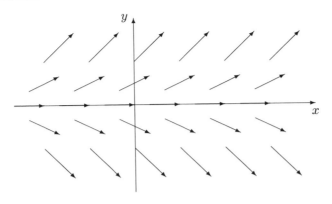

Figure 3.7 *Vector field* $v = (1, y)$.

along which the function u is constant. Indeed, let $c \in \mathbb{R}$ be an arbitrary constant. Then

$$\frac{\partial}{\partial x} u(x, y(x)) = \frac{\partial}{\partial x} u(x, ce^x) = u_x + u_y \underbrace{ce^x}_{y} = 0,$$

since u solves equation (3.12); further, we have

$$\frac{\partial}{\partial y} u(x, y(x)) = \frac{\partial}{\partial y} u(x, ce^x) = 0,$$

since none of the variables depend on y. The solution u is thus constant on every curve $y = ce^x$. The choice of the constant c then gives a particular curve, and the variables x and y are linked by the relation

$$ye^{-x} = c.$$

Thus, for an arbitrary differentiable function $f = f(z)$, the solution u can be written in the form

$$u(x, y) = f(c) = f(ye^{-x}).$$

$\square$

Example 3.5 We solve the problem

$$\begin{cases} u_x + yu_y = 0, \\ u(0, y) = y^2. \end{cases}$$

From the previous calculation, we have

$$u(x, y) = f(ye^{-x}).$$

However, also

$$y^2 = u(0, y) = f(y),$$

thus
$$u(x,y) = y^2 e^{-2x}.$$

□

Example 3.6 (Strauss [19]) Let us find all solutions of the equation
$$u_x + 2xy^2 u_y = 0.$$
Here $v = (1, 2xy^2)$ and the characteristics are the curves which solve the equation
$$\frac{dy}{dx} = 2xy^2.$$
Integrating, we find out that the functions
$$y = -\frac{1}{x^2 + c}, \quad c \in \mathbb{R},$$
and also
$$y \equiv 0$$
describe the required characteristic curves. See Figure 3.8. Similarly to Example 3.4, we obtain
$$c = -x^2 - \frac{1}{y},$$
and the solution can be written as
$$u(x,y) = \begin{cases} f\left(x^2 + \dfrac{1}{y}\right) & \text{for } y \neq 0, \\ \text{const.} & \text{for } y = 0. \end{cases}$$

In general, f can be an arbitrary differentiable function. However, in such a case, we have to understand the solution in the generalized sense. If we require u to be the classical solution, we would have to add more assumptions on f to ensure the continuity of u and its first derivatives at $y = 0$!

□

If we deal, in general, with the equation
$$\boxed{a(x,y)u_x(x,y) + b(x,y)u_y(x,y) = 0,}$$
we proceed quite analogously. Now, $v(x,y) = (a(x,y), b(x,y))$ and the characteristics are the curves which are given by the solution of the ODE
$$\frac{dy}{dx} = \frac{b(x,y)}{a(x,y)},$$

Section 3.3 Equations with Non-Constant Coefficients

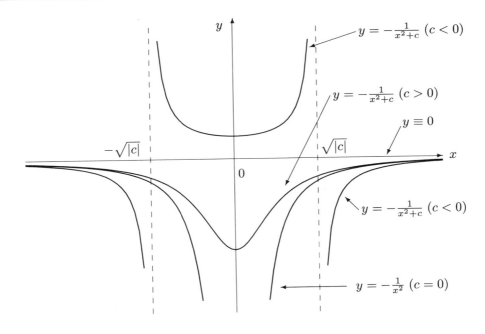

Figure 3.8 *Characteristics of the equation $u_x + 2xy^2 u_y = 0$.*

under the assumption $a(x,y) \neq 0$. The characteristics can be found also in the parametric form $x = x(t)$, $y = y(t)$. This parametric expression of the characteristics then solves the system
$$\begin{cases} \dot{x}(t) = a(x(t), y(t)), \\ \dot{y}(t) = b(x(t), y(t)). \end{cases}$$

Now, we return to equation (3.5) with constants a and b, and consider its more general version, namely,

$$au_x + bu_y + c(x,y)u = f(x,y),$$

where $c = c(x,y)$ is a nonconstant coefficient and $f = f(x,y)$ is a given function. The next example shows that, in that case, the coordinate method can be applied.

Example 3.7 Using the coordinate method, we find all solutions of the equation

(3.13) $\qquad u_x + 2u_y + (2x - y)u = 2x^2 + 3xy - 2y^2.$

Let us introduce the substitution

$$x' = ax + by = x + 2y, \quad y' = bx - ay = 2x - y.$$

According to the chain rule, equation (3.13) can be written in the form

$$5u_{x'} + y'u = x'y'.$$

This is nothing but a linear nonhomogeneous ODE of the first order with the variable x' and the parameter y'. First, we find the solution of the homogeneous equation

$$u_H(x', y') = f(y')e^{-\frac{1}{5}x'y'},$$

then, using, for instance, variation of parameters, we determine a particular solution of the nonhomogeneous equation

$$u_P(x', y') = x' - \frac{5}{y'}, \qquad y' \neq 0.$$

The general solution is then the sum of the homogeneous and particular solutions

$$u(x', y') = x' - \frac{5}{y'} + f(y')e^{-\frac{1}{5}x'y'}, \qquad y' \neq 0$$

and, after passing to the original variables x, y, we obtain

$$u(x, y) = x + 2y - \frac{5}{2x - y} + f(2x - y)e^{-\frac{1}{5}(2x^2 + 3xy - 2y^2)}, \qquad y \neq 2x.$$

If $y = 2x$, equation (3.13) reduces to $u_x + 2u_y = 0$, and thus $u(x, y) = \text{const.}$ on the line $y = 2x$.

The reader is invited to carry out the above steps in detail and to verify the correctness of the solution.

□

Also the method of characteristic coordinates can be applied to more general equations than (3.10), namely, to the case when $\lambda = \lambda(x, t)$ is not a constant in (3.10). The resulting equation (3.11) then has the form

$$u_\tau + \lambda(\xi + c\tau, \tau)u = f(\xi + c\tau, \tau).$$

The reader is invited to justify this fact.

3.4 Exercises

1. Solve the equation $u_t - 3u_x = 0$ with the initial condition $u(x, 0) = e^{-x^2}$ (see Figure 3.4).

$$[u(x, t) = e^{-(x+3t)^2}]$$

2. Solve the equation $u_t + 3u_x = 0$ with the initial condition $u(x, 0) = \sin x$ (see Figure 3.5).

$$[u(x, t) = \sin(x - 3t)]$$

Section 3.4 Exercises

3. Solve the equation $3u_y + u_{xy} = 0$ using substitution $v = u_y$.
$$[u(x,y) = e^{-3x}g(y) + f(x)]$$

4. Solve the linear equation $(1+x^2)u_x + u_y = 0$. Draw some characteristics.
$$[u(x,y) = f(y - \arctan x)]$$

5. Solve the equation $\sqrt{1-x^2}\,u_x + u_y = 0$ with the condition $u(0,y) = y$.
$$[u(x,y) = y - \arcsin x]$$

6. Using the coordinate method, solve the equation $au_x + bu_y + cu = 0$.
$$[u(x,y) = e^{-\frac{c}{a^2+b^2}} f(bx - ay)]$$

7. Using the coordinate method, solve the equation $u_x + u_y + u = e^{x+2y}$ with the initial condition $u(x,0) = 0$.
$$[u(x,y) = \tfrac{1}{4}\left(e^{x+2y} - e^{x-2y}\right)]$$

8. Solve the equation $u_t + au_x = x^2 t + 1$, where a is a constant, with the initial condition $u(x,0) = x + 2$.
$$[u(x,t) = x - at + 2 + t + \tfrac{x^2 t^2}{2} - \tfrac{1}{3}axt^3 + \tfrac{1}{12}a^2 t^4]$$

9. Solve the equation $u_t + t^\alpha u_x = 0$, where $\alpha > -1$ is a constant, with the initial condition $u(x,0) = \varphi(x)$.
$$[u(x,t) = \varphi\left(x - \tfrac{t^{\alpha+1}}{\alpha+1}\right)]$$

10. Solve the equation $u_t + xt\, u_x = x^2$ with the initial condition $u(x,0) = \varphi(x)$.
$$[u(x,t) = \varphi(xe^{-1/2t^2}) + x^2 e^{-t^2} \int_0^t e^{s^2} ds]$$

11. Find the general solution to the transport equation with decay $u_t + cu_x = -\lambda u$ using the transformation of independent variables
$$\xi = x - ct, \quad \tau = t.$$
$$[u(x,t) = e^{-\lambda t} f(x - ct)]$$

12. Show that the decay term in the transport equation with decay
$$u_t + cu_x = -\lambda u$$
can be eliminated by the substitution $w = ue^{\lambda t}$.

13. Solve the Cauchy problem
$$\begin{cases} u_t + u_x - 3u = t, & x \in \mathbb{R},\ t > 0, \\ u(x,0) = x^2, & x \in \mathbb{R}. \end{cases}$$
$$[u(x,t) = -\tfrac{1}{3}t - \tfrac{1}{9} + e^{3t}\left((x-t)^2 + \tfrac{1}{9}\right)]$$

14. Solve the transport equation with the convective term

$$u_t + 2u_x = -3u$$

under the condition $u(x, 0) = \frac{1}{1+x^2}$. (See Figure 3.9 and notice the influence of the convective term $3u$ on the solution on various time levels.

$$[u(x,t) = \frac{e^{-3t}}{1+4t^2-4tx+x^2}]$$

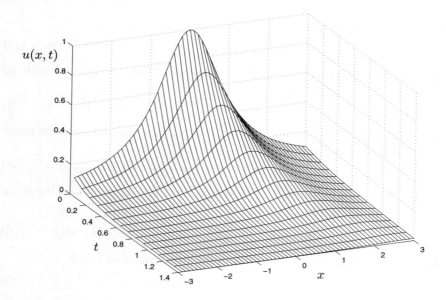

Figure 3.9 *Solution of the problem* $u_t + 2u_x = -3u$, $u(x,0) = 1/(1+x^2)$.

15. Solve the initial boundary value problem

$$\begin{cases} u_t + cu_x = -\lambda u, & x, t > 0, \\ u(x, 0) = 0, & x > 0, \quad u(0,t) = g(t), \quad t > 0. \end{cases}$$

Consider separately the cases $x > ct$ and $x < ct$. The boundary condition takes effect in the domain $x < ct$, whereas the initial condition influences the solution only in the domain $x > ct$.

$$[u(x,t) = g(t - \tfrac{x}{c})e^{-\tfrac{\lambda}{c}x} \text{ for } x - ct < 0, \quad u(x,t) = 0 \text{ for } x - ct > 0]$$

16. Find an implicit formula for the solution $u = u(x,t)$ of the initial value problem for the equation of transport reaction

$$\begin{cases} u_t + vu_x = -\frac{\alpha u}{\beta+u}, & x \in \mathbb{R},\ t > 0, \\ u(x, 0) = f(x), & x \in \mathbb{R}. \end{cases}$$

Here v, α, β are positive constants. Show that u can be always expressed in terms of x and t from the implicit formula.
$$[\beta \ln|u(x,t)| + u(x,t) = -\alpha t + f(x-vt) + \beta \ln|f(x-vt)|]$$

17. Find general solutions of the following equations.

    (a) $u_x + x^2 u_y = 0$,

    (b) $u_x + \sin x \, u_y = 0$,

    (c) $x \, u_x + y \, u_y = 0$,

    (d) $e^{x^2} u_x + x \, u_y = 0$,

    (e) $x \, u_x + y \, u_y = x^n$.

18. Solve the linear equation
$$x \, u_x - y \, u_y + y^2 u = y^2, \qquad x, y \neq 0.$$
$$[u(x,y) = f(xy)e^{y^2/2} + 1]$$

19. Consider the equation $y \, u_x - x \, u_y = 0$. Find initial conditions (initial curves) under which this problem has a unique solution, no solution, or infinitely many solutions.

    [a) $u_0(x,0) = x^2$, b) $u_0(x,y) = y$ on $x^2 + y^2 = 1$, c) $u_0(x,y) = 1$ on $x^2 + y^2 = 1$]

20. Consider the quasi-linear equation $u_y + a(u) u_x = 0$ with the initial condition $u(x,0) = h(x)$. Show that its solution can be given implicitly by $u = h(x - a(u)y)$. What are the characteristics? What happens if $a(h(x))$ is an increasing function?

21. Consider the equation

    (3.14) $$u_y = \left(\frac{y}{x} u\right)_x.$$

    Show that

    (a) the general solution of (3.14) is given by $u = x f(x^2 + y^2)$;

    (b) the function
    $$I(x,y) = \frac{x}{y} \int_0^\infty e^{-y\sqrt{1+t^2}} \cos xt \, dt$$
    satisfies equation (3.14);

    (c) the following identity is satisfied:
    $$\int_0^\infty e^{-y\sqrt{1+t^2}} \cos xt \, dt = \frac{y}{\sqrt{x^2+y^2}} \int_0^\infty e^{-\sqrt{(1+t^2)(x^2+y^2)}} dt, \quad y > 0.$$

22. Show that the initial value problem

$$u_t + u_x = 0, \quad u(x,t) = x \text{ on } x^2 + t^2 = 1$$

has no solution. However, if the initial data are given only over the semicircle that lies in the half-plane $x + t \leq 0$, a solution exists but is not differentiable along the characteristics coming from the two end points of the semicircle.

23. Show that the initial value problem

$$(t-x)u_x - (t+x)u_t = 0, \quad u(x,0) = f(x), \quad x > 0,$$

has no solution in general.

[Hint: The singular point at the origin gives rise to characteristics that spiral around the origin.]

24. Show that the equation $a(x)u_x + b(t)u_t = 0$ has the general solution $u(x,t) = F(A(x) - B(t))$, where $A'(x) = 1/a(x)$ and $B'(t) = 1/b(t)$.

25. Show that the equation $a(t)u_x + b(x)u_t = 0$ has the general solution $u(x,t) = F(B(x) - A(t))$, where $B'(x) = b(x)$ and $A'(t) = a(t)$.

26. Show that the initial value problem

$$u_t + u_x = x, \quad u(x,x) = 1$$

has no solution, and explain why.

27. Consider the semilinear equation

$$a(x,y,u(x,y))\,u_x + b(x,y,u(x,y))\,u_y = c(x,y,u(x,y))$$

and show that the method of characteristics yields

$$\frac{dx}{a(x,y,u(x,y))} = \frac{dy}{b(x,y,u(x,y))} = \frac{du(x,y)}{c(x,y,u(x,y))}.$$

[Hint: The differential equation can be understood as the scalar product of vectors $(a(x,y,u), b(x,y,u), c(x,y,u))$ and $(u_x, u_y, -1)$, where the last one represents the normal vector to the solution surface $u = u(x,y)$ in the Euclidean space (x,y,u).]

Exercises 3–7 are taken from Strauss [19], 8–10 from Barták et al. [4], 11, 13, 15, 16 from Logan [14], 18–21 from Stavroulakis and Tersian [18], 22–26 from Zauderer [22], and 27 from Keane [12].

4 Wave Equation in One Spatial Variable—Cauchy Problem in $\mathbb{R}$

4.1 String Vibrations and Wave Equation in One Dimension

The transport equation (see Section 3.1) was derived by a standard scheme: we substituted a particular constitutive law into the basic conservation law in its local form (1.8) and obtained the corresponding model. Now, we proceed in a little different way to derive another fundamental equation—the wave equation—that describes one of the most frequent phenomena in nature, the wave motion (let us recall electromagnetic waves, surface waves, or acoustic waves).

Let us consider a flexible string of length l and assume that there occur only small vibrations in the vertical direction (in the vertical plane). The displacement at a point x and time t will be denoted by a continuously differentiable function $u(x,t)$. The properties of the string are described by continuous functions $\rho(x,t)$ and $T(x,t)$ which represent the mass density and the inner tension of the string at a point x and time t. We assume that the tension T always acts in the direction tangent to the string profile at a point x. Now, let us consider an arbitrary but fixed string segment between points $x = a$, $x = b$ (see Figure 4.1). The angle formed by the tangent at a given point and the horizontal line will be a continuous function denoted by $\varphi(x,t)$. Let us notice that the relation

(4.1) $$\tan \varphi(x,t) = u_x(x,t)$$

holds.

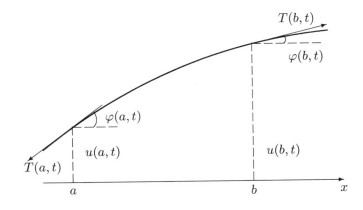

Figure 4.1 *String segment.*

To derive an equation describing the motion of the string, we use Newton's Second Law of Motion, which implies that the time change of the momentum in a given segment is equal to the acting force. We will assume that the only force which acts on the string segment is the tension caused by the neighboring parts of the string (gravity and damping are for now neglected). Since there is no movement in the horizontal direction, the following relation must hold:

(4.2) $$T(b,t)\cos\varphi(b,t) - T(a,t)\cos\varphi(a,t) = 0.$$

Since $[a,b]$ has been chosen arbitrarily, the quantity $T(x,t)\cos\varphi(x,t)$ does not depend on x. Let us denote

(4.3) $$\tau(t) = T(x,t)\cos\varphi(x,t).$$

In the vertical direction, the law of motion implies the kinetic equation

(4.4) $$\frac{d}{dt}\int_a^b \rho(x,t)u_t(x,t)\sqrt{1+u_x(x,t)^2}\,dx$$
$$= T(b,t)\sin\varphi(b,t) - T(a,t)\sin\varphi(a,t).$$

The reader should notice that this relation corresponds to the one-dimensional version of conservation law (1.3) with momentum as the quantity considered. To be able to pass to the local relation, we utilize another conservation law—the mass conservation law. It says that the time change of the total mass of a given segment is zero. (Here, the quantity considered is the mass, the flow as well as the sources are zero.) This, in other words, means that the mass of a given segment at an arbitrary time t must be equal to the mass of the same segment at time $t = 0$. If we denote $\rho_0(x) = \rho(x,0)$ and suppose that at the beginning, at time $t = 0$, the string is in its equilibrium state $u(x,0) \equiv u_0 = $ const., we obtain the equality

$$\int_a^b \rho(x,t)\sqrt{1+u_x(x,t)^2}\,dx = \int_a^b \rho_0(x)\,dx.$$

(The expression $\sqrt{1+u_x(x,t)^2}\,dx$ represents an element of arc length at time t, and it equals dx at time $t = 0$.) However, the interval $[a,b]$ has been chosen arbitrarily, thus the relation

(4.5) $$\rho(x,t)\sqrt{1+u_x(x,t)^2} = \rho_0(x)$$

must hold for any x and t.

Now, we can return to the kinetic equation (4.4). If we use (4.5) and change the order of differentiation and integration (which requires again some smoothness assumptions), we obtain

(4.6) $$\int_a^b \rho_0(x)u_{tt}(x,t)\,dx = T(b,t)\sin\varphi(b,t) - T(a,t)\sin\varphi(a,t).$$

Section 4.1 String Vibrations and Wave Equation in One Dimension

Employing (4.1), (4.2) and (4.3), we can rewrite the right-hand side in the following way:

$$T(b,t)\cos\varphi(b,t)\tan\varphi(b,t) - T(a,t)\cos\varphi(a,t)\tan\varphi(a,t)$$
$$= \tau(t)(u_x(b,t) - u_x(a,t)) = \tau(t)\int_a^b u_{xx}(x,t)\,dx.$$

Hence, from (4.6) we obtain

$$\int_a^b \rho_0(x) u_{tt}(x,t)\,dx = \tau(t)\int_a^b u_{xx}(x,t)\,dx.$$

This relation must hold again for an arbitrary interval $[a,b]$, thus we get the differential formulation

$$\rho_0(x) u_{tt}(x,t) = \tau(t) u_{xx}(x,t).$$

In the special case that $\rho_0(x) \equiv \rho_0$, $\tau(t) \equiv \tau_0$, and if we denote $c = \sqrt{\tau_0/\rho_0}$, we arrive at the fundamental equation of mathematical modeling describing string vibrations in one dimension:

(4.7)
$$\boxed{u_{tt} = c^2 u_{xx}.}$$

Equation (4.7) is called the *wave equation* and the constant $c > 0$ expresses the *speed of wave propagation*.

The basic wave equation can be modified in various ways.

(i) If the presence of *external damping* (for example, resistance of the surrounding medium) is taken into account, the equation is enriched by a term proportional to the velocity u_t:

$$u_{tt} - c^2 u_{xx} + r u_t = 0$$

with the *damping* coefficient $r > 0$.

(ii) If we consider an *elastic response* of the surrounding medium in the model, we obtain

$$u_{tt} - c^2 u_{xx} + k u = 0$$

with the *stiffness* $k > 0$.

(iii) If we want to include the presence of an *external force* in the model, we obtain a *nonhomogeneous* wave equation

$$u_{tt} - c^2 u_{xx} = f(x,t).$$

(iv) All these effects can appear simultaneously. Then we obtain an equation of the form

$$u_{tt} - c^2 u_{xx} + r u_t + k u = f(x,t).$$

4.2 Cauchy Problem on the Real Line

First, let us consider the wave equation

(4.8) $$u_{tt} = c^2 u_{xx}, \quad x \in \mathbb{R},\ t > 0$$

and look for its general solution. For illustration, we can imagine an "infinitely long" string. We will present two methods of finding the general solution of equation (4.8). Both methods are standard and the reader can find them in many other textbooks.

Method I (transformation to a system of two first order equations)

Equation (4.8) can be formally rewritten as

(4.9) $$\left(\frac{\partial}{\partial t} - c\frac{\partial}{\partial x}\right)\left(\frac{\partial}{\partial t} + c\frac{\partial}{\partial x}\right) u = 0.$$

If we introduce a new function v by the relation $v = u_t + cu_x$, we transform the original equation (4.8) into a system of two equations of the first order

(4.10) $$\begin{cases} v_t - cv_x = 0, \\ u_t + cu_x = v. \end{cases}$$

Both the equations are now solvable by methods introduced in Section 3.2. The first equality implies

$$v(x,t) = h(x + ct),$$

where h is an arbitrary differentiable function. We substitute for v into the latter equation of system (4.10) and obtain

$$u_t + cu_x = h(x + ct),$$

which represents a transport equation with constant coefficients and non-zero right-hand side. The solution of the corresponding homogeneous equation has the form

(4.11) $$u_H(x,t) = g(x - ct).$$

Since the right-hand side of the equation is formed by an arbitrary differentiable function of the argument $x + ct$, one particular solution (which reflects the influence of the right-hand side) must be also an arbitrary differentiable function of the same argument, thus

(4.12) $$u_P(x,t) = f(x + ct).$$

The general solution of the wave equation is then the sum of solutions (4.11) and (4.12):

(4.13) $$\boxed{u(x,t) = f(x + ct) + g(x - ct).}$$

Method II (the method of characteristics)

The second way of derivation of the general solution of the wave equation on the real line consists in introducing special coordinates

$$\xi = x + ct, \quad \eta = x - ct.$$

According to the chain rule, we have

$$\partial_x = \partial_\xi + \partial_\eta, \qquad \partial_t = c\partial_\xi - c\partial_\eta$$

and thus

$$\partial_t - c\partial_x = -2c\partial_\eta,$$
$$\partial_t + c\partial_x = 2c\partial_\xi.$$

The reader is invited to verify these formulas. After substituting it into the expression (4.9), we obtain a transformed equation

$$-4c^2 \partial_\eta \partial_\xi u = 0$$

or, equivalently,

$$u_{\xi\eta} = 0.$$

Its solution has been found in Chapter 2 (see Example 2.3), namely

$$u(\xi, \eta) = f(\xi) + g(\eta),$$

where f and g are again arbitrary differentiable functions. If we go back to the original variables x and t, we obtain the foregoing general solution of the one-dimensional wave equation

$$u(x,t) = f(x + ct) + g(x - ct).$$

As we can see, the solution is the sum of two traveling waves (cf. Section 3.1), the left one and the right one, which move at the speed $c > 0$. Lines $x+ct = \text{const.}$, $x-ct = \text{const.}$, along which the traveling waves propagate, are called the *characteristics of the wave equation*.

If we now consider an initial value (Cauchy) problem

(4.14)
$$\begin{array}{l} u_{tt} = c^2 u_{xx}, \quad x \in \mathbb{R}, \; t > 0, \\ u(x,0) = \varphi(x), \; u_t(x,0) = \psi(x), \end{array}$$

the particular forms of functions f and g can be determined in terms of the given functions φ and ψ, which describe the *initial displacement* and the *initial velocity* of the searched wave.

If we start with the general solution (4.13), then the following equalities hold for $t = 0$:

$$\varphi(x) = f(x) + g(x), \qquad \psi(x) = cf'(x) - cg'(x).$$

The first equality implies (assuming that all indicated derivatives exist)

$$\varphi'(x) = f'(x) + g'(x),$$

which, in combination with the latter equality, gives

$$f'(x) = \frac{1}{2}\varphi'(x) + \frac{1}{2c}\psi(x),$$

$$g'(x) = \frac{1}{2}\varphi'(x) - \frac{1}{2c}\psi(x)$$

and, after integration,

$$f(x) = \frac{1}{2}\varphi(x) + \frac{1}{2c}\int_0^x \psi(\tau)\,d\tau + A,$$

$$g(x) = \frac{1}{2}\varphi(x) - \frac{1}{2c}\int_0^x \psi(\tau)\,d\tau + B,$$

where A, B are integration constants. The condition $u(x, 0) = \varphi(x)$, however, implies $A + B = 0$. After substituting the previous relations into the general expression of solution (4.13), we obtain the solution of the Cauchy problem for the wave equation in one dimension:

(4.15) $$u(x,t) = \frac{1}{2}[\varphi(x+ct) + \varphi(x-ct)] + \frac{1}{2c}\int_{x-ct}^{x+ct}\psi(\tau)d\tau.$$

This formula was derived by d'Alembert in 1746. The first term on the right-hand side expresses the influence of the initial displacement: the initial wave is divided into two parts, the former proceeding in the direction of the negative x half-axis at a speed c, and the latter proceeding in the direction of the positive x half-axis at the same speed c. The integral on the right-hand side expresses the influence of the initial velocity. The solution expressed by d'Alembert's formula is determined uniquely (see Exercise 6 in Section 10.9).

The following assertion is the basic existence and uniqueness result for the wave equation.

Theorem 4.1 *Let* $\varphi \in C^2$, $\psi \in C^1$ *on the entire real line* $\mathbb{R}$. *The Cauchy problem (4.14) for the wave equation on the real line with the initial displacement* $\varphi(x)$ *and the initial velocity* $\psi(x)$ *has a unique classical solution* $u \in C^2$ *given by d'Alembert's formula (4.15).*

Example 4.2 We find a solution of the wave equation on the real line, if the initial displacement is zero and the initial velocity is given by $\sin x$.

Section 4.2 Cauchy Problem on the Real Line

This problem can be written as a Cauchy problem

(4.16)
$$\begin{cases} u_{tt} = c^2 u_{xx}, & x \in \mathbb{R},\ t > 0, \\ u(x,0) = 0,\ u_t(x,0) = \sin x. \end{cases}$$

After substituting the initial conditions into d'Alembert's formula (4.15), we obtain

$$u(x,t) = \frac{1}{2c} \int_{x-ct}^{x+ct} \sin \tau \, d\tau = \frac{1}{2c}[\cos(x-ct) - \cos(x+ct)]$$

and, applying the trigonometric formula $\cos\alpha - \cos\beta = -2\sin\frac{\alpha-\beta}{2}\sin\frac{\alpha+\beta}{2}$, the solution $u(x,t)$ can be written in the form

(4.17)
$$u(x,t) = \frac{1}{c}\sin ct \sin x.$$

□

Let us note the following special feature of solution (4.17): the zero points of u lie at the points $x = k\pi$, $k \in \mathbb{Z}$, for arbitrary $t \geq 0$. They do not "travel" along the x-axis with growing time. Solutions of the wave equation with the above mentioned property are called *standing waves* (see Figure 4.2).

Remark 4.3 In fact, a more general assertion than Theorem 4.1 holds. Namely, the initial value problem for the wave equation has a unique classical solution if and only if $\varphi \in C^2$ and $\psi \in C^1$. These assumptions on initial conditions are, however, strongly restrictive and very often contradict the practical problems that necessarily have to be solved by methods of mathematical modeling. These problems, if they are understood in the sense of a classical solution, are ill-posed. This fact caused big difficulties to mathematicians in the eighteenth century and it took a long time before they came to a more general notion of a solution (weak solution, very weak solution, generalized solution, strong solution, ...). In this text, we will not deal with these questions in detail. We restrict ourselves only to the statement that formula (4.15), expressing a solution of the initial value problem for the wave equation in explicit form, makes sense also for much more general initial conditions than $\varphi \in C^2$, $\psi \in C^1$. The solution u can be then viewed, for instance, as a function which satisfies the differential equation only at those points where the corresponding partial derivatives u_{tt} and u_{xx} exist. On the other hand, the set of points where these partial derivatives do not exist (and hence the equation itself does not make sense) cannot be "too large".

If φ is a C^2 function and ψ is a C^1 function on $\mathbb{R}$ with the exception of a finite number of points (the so called *singular points* or *singularities*) then (4.15) makes sense, the partial derivatives of u exist and are continuous with the exception of a finite number of lines in the xt plane and equation (4.8) holds at every point which does not belong to these lines. In such a way, we will understand also solutions of the following Cauchy problems. However, the reader should notice that even more general functions

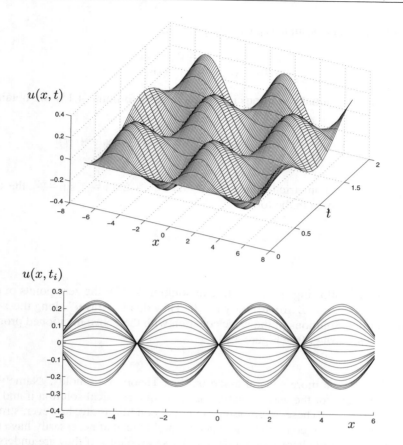

Figure 4.2 *Standing waves—a solution of the initial value problem (4.16) with $c = 4$.*

φ and ψ can be considered (for example, locally integrable) and the corresponding solution of (4.8) makes sense if it is understood in a more general sense. Existence and uniqueness results can be still proved in such a more general setting.

Example 4.4 (Strauss [19]) We solve the wave equation with the initial displacement

$$\varphi(x) = \begin{cases} b - \frac{b}{a}|x| & \text{for } |x| \leq a, \\ 0 & \text{for } |x| > a \end{cases}$$

and with zero initial velocity $\psi(x) \equiv 0$.

This problem describes the behavior of an infinitely long string which is at time $t = 0$ displaced by "three fingers" and then released. The three points $x = -a, 0, a$ represent singularities of the initial displacement φ. According to d'Alembert's for-

Section 4.2 Cauchy Problem on the Real Line

mula (4.15), the corresponding solution has the form

$$u(x,t) = \frac{1}{2}\left[\varphi(x-ct) + \varphi(x+ct)\right].$$

It is a sum of two "triangle functions" which diverge with increasing time. The shape of the solution on particular time levels is sketched in Figure 4.3, the whole graph of the function $u(x,t)$ is illustrated by Figure 4.4 (for the values $c = 2$, $b = 1$, $a = 2$). The reader should observe the lines in the xt plane where the partial derivatives of u do not exist and thus equation (4.8) does not make sense.

□

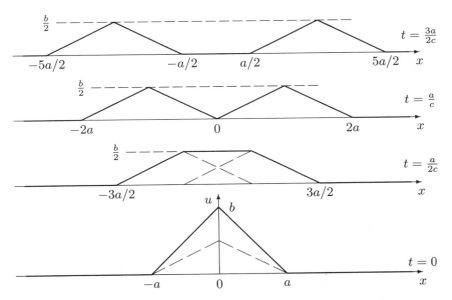

Figure 4.3 *Solution of Example 4.4 on particular time levels.*

Example 4.5 (Strauss [19]) Let us solve the wave equation with zero initial displacement $\varphi(x) \equiv 0$ and with the initial velocity

$$\psi(x) = \begin{cases} 1 & \text{for } |x| \leq a, \\ 0 & \text{for } |x| > a. \end{cases}$$

This problem can be regarded as a simplified model of the behavior of an infinitely long string after a stroke by a hammer of width $2a$. Here, the two points $x = -a, a$

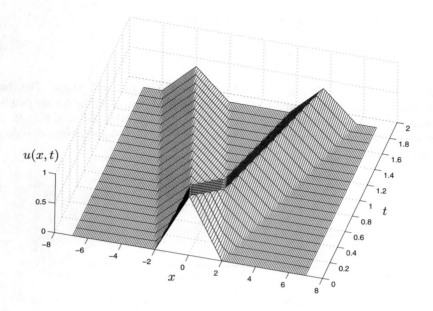

Figure 4.4 *Graph of solution from Example 4.4 for $c = 2$, $b = 1$, $a = 2$.*

represent singularities of the initial velocity ψ. D'Alembert's formula implies

$$u(x,t) = \frac{1}{2c} \int_{x-ct}^{x+ct} \psi(\tau)\,d\tau = \frac{1}{2c} \times \text{ length of interval } \{(-a,a) \cap (x-ct, x+ct)\}.$$

The shape of the solution on particular time levels is sketched in Figure 4.5, the whole graph of the function $u(x,t)$ is illustrated by Figure 4.6, where the values of parameters are chosen as $c = 2.3$ and $a = 1.3$. The reader is asked to pay attention to the lines in the xt plane where the partial derivatives of u do not exist and thus equation (4.8) does not make sense.

□

Section 4.2 Cauchy Problem on the Real Line

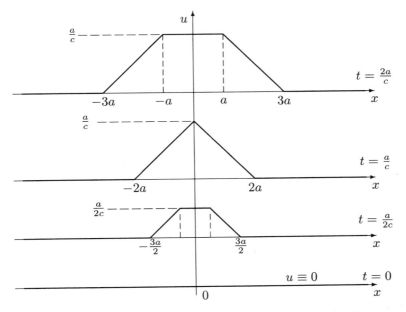

Figure 4.5 *Solution of Example 4.5 on particular time levels.*

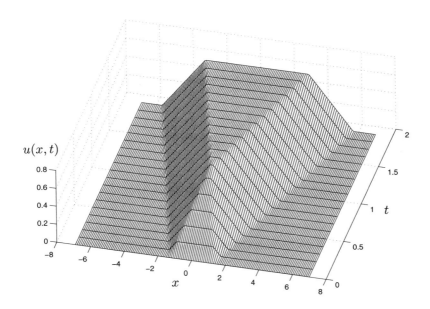

Figure 4.6 *Graph of solution from Example 4.5 for $c = 2.3$, $a = 1.3$.*

Principle of Causality. Let us investigate the solution of the initial value problem for the wave equation on the real line in more detail. We find out that the initial condition at the point $(x_0, 0)$ can "spread" only to that part of the xt plane which lies between the lines with equations $x \pm ct = x_0$ (the characteristics passing through the point $(x_0, 0)$). See Figure 4.7. The sector with these boundary points is called the *domain of influence* of the point $(x_0, 0)$.

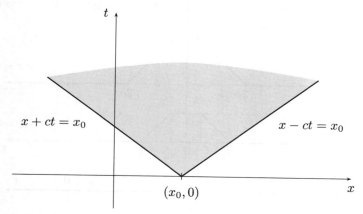

Figure 4.7 *Domain of influence of the point $(x_0, 0)$ at time $t \geq 0$.*

In particular, this means that the initial conditions with the property

$$\varphi(x) = \psi(x) \equiv 0 \quad \text{for } |x| > R$$

result in the solution which is identically zero "to the right" of the line $x - ct = R$ and "to the left" of the line $x + ct = -R$ (see Figure 4.8).

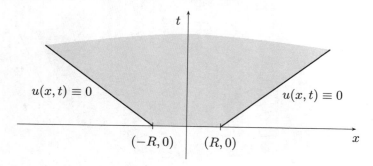

Figure 4.8 *Domain of influence of the interval $(-R, R)$ at time $t \geq 0$.*

The opposite (dual) view of the above situation is the following: let us choose an arbitrary point (x, t) and ask what values of the initial conditions on the x-axis (for

$t = 0$) can influence the value of the solution at a point (x, t). The above-mentioned information implies that these are just the values $\varphi(x)$, $\psi(x)$ for x from the interval between $x - ct$ and $x + ct$ (see Figure 4.9).

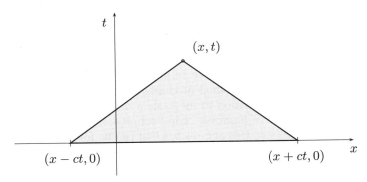

Figure 4.9 *Domain of dependence of the point (x, t).*

The triangle $\triangle_{xt}$ with vertices at the points $(x - ct, 0)$, $(x + ct, 0)$ and (x, t) is called the *domain of dependence* (or the *characteristic triangle*) of the point (x, t).

4.3 Wave Equation with Sources

Let us now consider the Cauchy problem for the wave equation with a non-zero right-hand side

(4.18)
$$\boxed{\begin{aligned} u_{tt} - c^2 u_{xx} &= f(x, t), \quad x \in \mathbb{R}, \ t > 0, \\ u(x, 0) &= \varphi(x), \ u_t(x, 0) = \psi(x). \end{aligned}}$$

The following existence and uniqueness result generalizes Theorem 4.1.

Theorem 4.6 *Let $\varphi \in C^2$, $\psi \in C^1$, $f \in C^1$. The initial value problem (4.18) has a unique classical solution which has the form*

(4.19)
$$\boxed{\begin{aligned} u(x, t) &= \frac{1}{2} [\varphi(x + ct) + \varphi(x - ct)] \\ &\quad + \frac{1}{2c} \int_{x-ct}^{x+ct} \psi(y) \, dy + \frac{1}{2c} \iint_{\triangle} f(y, s) \, dy \, ds. \end{aligned}}$$

The symbol $\triangle = \triangle_{xt}$ in the last integral represents the characteristic triangle, i.e., the domain of dependence of the point (x, t) up to the time $t = 0$:

$$\iint_{\triangle} f(y, s) \, dy \, ds = \int_0^t \int_{x-c(t-s)}^{x+c(t-s)} f(y, s) \, dy \, ds.$$

Notice that the principle of causality holds again even for the wave equation with external forces!

Remark 4.7 The reader can easily check that the function given by (4.19) is indeed a classical solution of problem (4.18). (See Exercise 14 in Section 4.4.) However, the reader should notice that the classical solution exists under the more general assumption $f \in C$.

There are several ways how to derive formula (4.19). One of the possibilities is based on the application of the method of characteristics, and another one uses the transformation of the wave equation to the system of two transport equations. The latter derivation is the scope of Exercise 21 in Section 4.4. In what follows, we focus our attention on other two standard approaches that can be found in several textbooks and have not been mentioned yet, namely, the use of *Green's Theorem* and application of the *Operator Method* (cf. Strauss [19]).

Use of Green's Theorem. For simplicity, let us consider a fixed point (x_0, t_0) and assume that u is a classical solution of (4.18). We integrate the wave equation over the domain of dependence of the point (x_0, t_0), that is, over the characteristic triangle $\triangle$:

$$\iint_\triangle f \, dx \, dt = \iint_\triangle (u_{tt} - c^2 u_{xx}) \, dx \, dt.$$

Now, we apply Green's Theorem to the right-hand side. It reads:

$$\iint_\triangle (P_x - Q_t) \, dx \, dt = \int_{\partial \triangle} P \, dt + Q \, dx$$

for arbitrary continuously differentiable functions P, Q. The curve integral over the boundary $\partial \triangle$ of the domain $\triangle$ is considered in the positive direction, that is in the counterclockwise direction. In our case, we set $P = -c^2 u_x$, $Q = -u_t$. If we denote the particular sides of the characteristic triangle by L_0, L_1, L_2 (see Figure 4.10), we obtain

$$\iint_\triangle f \, dx \, dt = \int_{L_0 \cup L_1 \cup L_2} -c^2 u_x \, dt - u_t \, dx,$$

which can be written as a sum of three curve integrals over the corresponding straight line segments.

On the side L_0, we have $t = 0$, $dt = 0$ and $u_t(x, 0) = \psi(x)$, thus

$$\int_{L_0} -c^2 u_x \, dt - u_t \, dx = -\int_{x_0 - ct_0}^{x_0 + ct_0} \psi(x) \, dx.$$

On L_1, we have $x + ct = x_0 + ct_0$ and thus $dx + c \, dt = 0$. Hence, we obtain

$$\int_{L_1} -c^2 u_x \, dt - u_t \, dx = c \int_{L_1} u_x \, dx + u_t \, dt = c \int_{L_1} du = cu(x_0, t_0) - c\varphi(x_0 + ct_0).$$

Section 4.3 Wave Equation with Sources

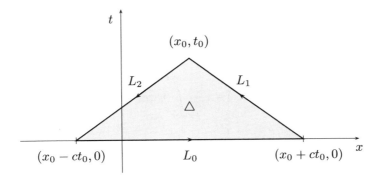

Figure 4.10 *Characteristic triangle of the point (x_0, t_0).*

Similarly, for L_2 ($dx - c\,dt = 0$),

$$\int_{L_2} -c^2 u_x \, dt - u_t \, dx = -c \int_{L_2} u_x \, dx + u_t \, dt = -c \int_{L_2} du = -c\varphi(x_0 - ct_0) + cu(x_0, t_0).$$

Combining these three partial results, we obtain

$$\iint_\Delta f \, dx \, dt = 2cu(x_0, t_0) - c[\varphi(x_0 + ct_0) + \varphi(x_0 - ct_0)] - \int_{x_0 - ct_0}^{x_0 + ct_0} \psi(x) \, dx$$

wherefrom the required form of the solution u at the point (x_0, t_0) follows.

Operator Method. This time we try to derive the solution of the initial value problem for the nonhomogeneous wave equation on the basis of an analogue of the solution of ODE

(4.20) $$\frac{d^2 v}{dt^2} + A^2 v(t) = f(t)$$

with initial conditions

$$v(0) = \varphi, \qquad \frac{dv}{dt}(0) = \psi,$$

where φ and ψ are real numbers. Applying the variation of constants formula, the solution of equation (4.20) for a constant $A \neq 0$ can be written in the form

(4.21) $$v(t) = S'(t)\varphi + S(t)\psi + \int_0^t S(t - s)f(s)\,ds,$$

where

(4.22) $$S(t) = \frac{1}{A} \sin At, \qquad S'(t) = \cos At.$$

Thus, in the case $\varphi = 0$, $f = 0$, the solution reduces to $v(t) = S(t)\psi$.

Now, we turn back to our wave equation. We have derived that the solution of the homogeneous equation for $\varphi(x) \equiv 0$, $f(x,t) \equiv 0$ can be written in the form

$$u(x,t) = \frac{1}{2c} \int_{x-ct}^{x+ct} \psi(y)\, dy.$$

If we define the source operator $S(t)$ by

(4.23)
$$S(t)\psi(x) = \frac{1}{2c} \int_{x-ct}^{x+ct} \psi(y)\, dy,$$

we can write

$$u(x,t) = S(t)\psi(x).$$

Analogously to the first term on the right-hand side of relation (4.21), we could expect the reaction on the non-zero initial displacement in the form $\frac{\partial}{\partial t} S(t)\varphi$. Indeed, we have

(4.24)
$$\frac{\partial}{\partial t} S(t)\varphi(x) = \frac{\partial}{\partial t} \frac{1}{2c} \int_{x-ct}^{x+ct} \varphi(y)\, dy = \frac{1}{2c}[c\varphi(x+ct) - (-c)\varphi(x-ct)],$$

which corresponds to d'Alembert's formula (4.15).

Now, let us consider only the influence of the right-hand side. To this end, put $\varphi = \psi = 0$. If we use again the analogue of the solution of the ODE (4.21), we write the corresponding solution of the wave equation in the form

$$u(x,t) = \int_0^t S(t-s) f(x,s)\, ds;$$

thus, using the definition of $S(t)$ in (4.23), we conclude that

(4.25)
$$u(x,t) = \int_0^t \left[\frac{1}{2c} \int_{x-c(t-s)}^{x+c(t-s)} f(y,s)\, dy \right] ds = \frac{1}{2c} \iint_\Delta f(y,s)\, dy\, ds.$$

Putting together (4.23)–(4.25), we arrive at (4.19).

The approach based on the idea that the knowledge of a solution of the homogeneous equation can be used for the derivation of a solution of the nonhomogeneous equation is, in connection with the wave equation, called *Duhamel's principle*.

4.4 Exercises

1. Derive the model of the vibrating string including the gravitation force:
$$u_{tt} = c^2 u_{xx} - g,$$
here g is the constant representing the gravitational acceleration. Notice that gravity acts at every point of the string in the vertical direction, that is, it appears as the source term in the momentum conservation law.

2. Derive a *damped wave equation*
$$u_{tt} = c^2 u_{xx} - k u_t$$
describing vibrations of the string whose vertical motion is decelerated by a damping force proportional to the string velocity. Here, k is a damping coefficient.

3. Verify that the function
$$u(x,t) = \frac{1}{2c} \int_{x-ct}^{x+ct} g(\xi)\, d\xi$$
solves the wave equation $u_{tt} = c^2 u_{xx}$, where c is a constant and g is a continuously differentiable function. Use the rule for the derivative of the integral with respect to parameters x and t occurring in the limits of integration.

4. A linear approximation of the one-dimensional isotropic flow of an ideal gas is given by
$$u_t + \rho_x = 0, \quad u_x + c^2 \rho_t = 0,$$
where $u = u(x,t)$ is the velocity of the gas and $\rho = \rho(x,t)$ is its density. Show that u and ρ satisfy the wave equation.

5. Deriving the general solution (4.13) of the wave equation, we have used the fact that the linear wave operator $L = (\partial_t)^2 - c^2 (\partial_x)^2$ is *reducible* (or *factorable*), that is, it can be written as a product of linear first-order operators: $L = L_1 L_2$. Using the same idea, find the general solutions of the following equations.

   (a) $u_{xx} + u_x = u_{yy} + u_y$.
   [$L_1 = \partial_x - \partial_y$, $L_2 = \partial_x + \partial_y + 1$. The general solution can be written as $u(x,y) = \varphi(x+y) + e^{-x}\psi(x-y)$ or as $u(x,y) = \varphi(x+y) + e^{-y}h(x-y)$, where φ, ψ and h are arbitrary differentiable functions.]

   (b) $3u_{xx} + 10u_{xy} + 3u_{yy} = 0$.
   [$L_1 = 3\partial_x + \partial_y$, $L_2 = \partial_x + 3\partial_y$; $u(x,y) = \varphi(3x - y) + \psi(x - 3y)$ with arbitrary functions φ, ψ]

6. Solve the Cauchy problem $u_{tt} = c^2 u_{xx}$, $u(x,0) = e^x$, $u_t(x,0) = \sin x$.
$$[u(x,t) = \tfrac{1}{2}(e^{x+ct} + e^{x-ct}) - \tfrac{1}{2c}(\cos(x+ct) - \cos(x-ct))]$$

7. Solve the Cauchy problem $u_{tt} = c^2 u_{xx}$, $u(x,0) = \ln(1+x^2)$, $u_t(x,0) = 4+x$.

$$[u(x,t) = \ln\sqrt{(1+(x+ct)^2)(1+(x-ct)^2)} + t(4+x)]$$

8. Solve the Cauchy problem $u_{tt} - 3u_{xt} - 4u_{xx} = 0$, $u(x,0) = x^2$, $u_t(x,0) = e^x$. Proceed in the same way as when deriving the general solution of the wave equation.

$$[u(x,t) = x^2 + 4t^2 + \tfrac{e^{x+4t} - e^{x-t}}{5}]$$

9. Solve the Cauchy problem $u_{tt} - u_{xx} = 0$, $u(x,0) = 0$, $u_t(x,0) = -2xe^{-x^2}$.

$$[u(x,t) = \tfrac{1}{2}(e^{-(x+t)^2} - e^{-(x-t)^2})]$$

10. Solve the Cauchy problem $u_{tt} - u_{xx} = 0$, $u(x,0) = 0$, $u_t(x,0) = \tfrac{x}{(1+x^2)^2}$.

$$[u(x,t) = \tfrac{1}{4}(\tfrac{1}{1+(x-t)^2} - \tfrac{1}{1+(x+t)^2})]$$

11. Solve the Cauchy problem $u_{tt} - u_{xx} = 0$ for

$$u(x,0) = \begin{cases} e^{-x}, & |x| < 1, \\ 0, & |x| > 1, \end{cases} \quad u_t(x,0) = 0.$$

12. Solve the Cauchy problem $u_{tt} - u_{xx} = 0$ for

$$u(x,0) = 0, \quad u_t(x,0) = \begin{cases} e^{-x}, & |x| < 1, \\ 0, & |x| > 1. \end{cases}$$

13. Prove that the function

$$u(x,t) = \frac{1}{2}\left[e^{-(x-2t)^2} + e^{-(x+2t)^2}\right]$$

(see Figure 4.11) solves the Cauchy problem

(4.26) $\quad \begin{cases} u_{tt} - 4u_{xx} = 0, & x \in \mathbb{R},\ t > 0, \\ u(x,0) = e^{-x^2}, & u_t(x,0) = 0. \end{cases}$

14. By a simple substitution, verify that the function

$$u(x,t) = \frac{1}{2c}\iint_\Delta f(y,s)\,dy\,ds$$

solves the nonhomogeneous wave equation $u_{tt} - c^2 u_{xx} = f$ (cf. Theorem 4.6). Explain why we need the assumption $f \in C^1$.

15. Solve the Cauchy problem $u_{tt} = c^2 u_{xx} + xt$, $u(x,0) = 0$, $u_t(x,0) = 0$.

$$[u(x,t) = \tfrac{xt^3}{6}]$$

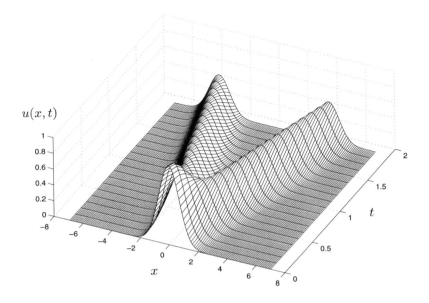

Figure 4.11 *Solution of problem (4.26).*

16. Solve the Cauchy problem $u_{tt} = c^2 u_{xx} + e^{at}$, $u(x,0) = 0$, $u_t(x,0) = 0$.

$$[u(x,t) = \tfrac{1}{a^2}(e^{at} - at - 1)]$$

17. Solve the Cauchy problem $u_{tt} = c^2 u_{xx} + \cos x$, $u(x,0) = \sin x$, $u_t(x,0) = 1+x$.

$$[u(x,t) = \cos ct \sin x + (1+x)t + \tfrac{\cos x}{c^2} - \tfrac{\cos x \cos ct}{c^2}]$$

18. Solve the Cauchy problem $u_{tt} - u_{xx} = e^{x-t}$, $u(x,0) = 0$, $u_t(x,0) = 0$.

$$[u(x,t) = \tfrac{1}{4}(e^{x+t} - e^{x-t}) - \tfrac{1}{2}te^{x-t}]$$

19. Solve the Cauchy problem $u_{tt} - u_{xx} = \sin x$, $u(x,0) = \cos x$, $u_t(x,0) = x$.

$$[u(x,t) = \cos x \cos t + xt + \sin x - \sin x \cos t]$$

20. Solve the Cauchy problem $u_{tt} - u_{xx} = x^2$, $u(x,0) = \cos x$, $u_t(x,0) = 0$.

$$[u(x,t) = \cos x \cos t + \tfrac{x^2 t^2}{2} + \tfrac{t^4}{12}]$$

21. Derive the solution of the nonhomogeneous wave equation in another possible way:

(a) Rewrite the equation to the system

$$u_t + cu_x = v, \quad v_t - cv_x = f.$$

(b) In the case of the former equation, find a solution u dependent on v in the form

$$u(x,t) = \int_0^t v(x - ct + cs, s)\,ds.$$

(c) Similarly, solve the latter equation, i.e., find v dependent on f.
(d) Insert the result of part (c) into the result of part (b).

22. Consider the *telegraph equation* $u_{xx} - \frac{1}{c^2}u_{tt} + \alpha u_t + \beta u = 0$ and put $v = u$, $w = u_x$ and $z = u_t$. Show that v, w and z satisfy the following system of three equations:

$$v_t - z = 0,$$
$$w_t - z_x = 0,$$
$$z_t - c^2(w_x + \alpha z + \beta v) = 0.$$

Exercises 1, 2, 13 are taken from Logan [14], 4, 11, 12 from Keane [12], 5, 18–20 from Stavroulakis and Tersian [18], 6–8, 15, 16, 21 from Strauss [19], and 9, 10 from Asmar [3].

5 Diffusion Equation in One Spatial Variable—Cauchy Problem in $\mathbb{R}$

5.1 Diffusion and Heat Equations in One Dimension

Diffusion. Let us study the behavior of a gas in a one-dimensional tube. We denote its concentration at a point x and time t by $u = u(x,t)$ (the state function) and the corresponding flux density by $\phi = \phi(x,t)$ (the flow function). If we do not admit any sources, then these two quantities obey the one-dimensional conservation law (1.8)

$$u_t + \phi_x = 0.$$

Experiments show that the molecules of the gas move from the higher concentration area to the lower concentration area, and that the higher is the concentration gradient, the greater is the flux density. The simplest relation (the constitutive law) that corresponds to these assumptions is the linear dependence

(5.1) $$\phi = -ku_x,$$

where k is a constant of proportionality. The minus sign ensures that if $u_x < 0$, then ϕ is positive and the flow moves "to the right". Equation (5.1) is called *Fick's First Law of diffusion* and k is the *diffusion constant*. If we insert (5.1) into the conservation law, we obtain the one-dimensional *diffusion equation*

(5.2) $$\boxed{u_t - ku_{xx} = 0}$$

which expresses *Fick's Second Law of diffusion*.

Heat Equation. The same assumptions as those we have used in the derivation of the diffusion equation can be applied also in modeling the heat flow. Let us consider a one-dimensional bar with constant mass density ρ and constant specific heat capacity c. If we denote by $u = u(x,t)$ the thermodynamic temperature at a point x and time t, then the quantity $\rho c u(x,t)$ represents the volume density of internal energy. In this case, the conservation law (called the *heat conservation law*) expresses the balance between the internal energy $\rho c u$ and the heat flux ϕ (for simplicity, we admit no sources):

(5.3) $$(\rho c u)_t(x,t) + \phi_x(x,t) = 0.$$

The constitutive law connecting the density of the heat flux ϕ and the temperature u is *Fourier's heat law* which says that the density of the heat flux is directly proportional to the temperature gradient with a negative constant of proportionality:

$$\phi = -Ku_x.$$

The constant K represents the *heat* (or *thermal*) *conductivity*. Fourier's law is an equivalent of Fick's first law: heat flows from warmer places of the body to colder places. If we substitute for the heat flux back into (5.3), we obtain

(5.4) $$u_t - k u_{xx} = 0, \qquad k := \frac{K}{\rho c},$$

which is again a diffusion equation in one dimension. (Here, the constant k is called the *thermal diffusivity*.) Both phenomena, the heat flow and the diffusion, can be thus modeled by the same equation.

Transport with Diffusion. If we want to include the transport with diffusion into the model, then the flux density must satisfy the constitutive law

$$\phi = cu - k u_x,$$

and using the conservation law we obtain the equation

(5.5) $$u_t + c u_x - k u_{xx} = 0.$$

In this way we can describe, for instance, the density distribution of some chemical which is drifted by a flowing liquid at a speed c and, at the same time, diffuses into this liquid with a diffusion constant k.

5.2 Cauchy Problem on the Real Line

Let us consider a Cauchy problem for the diffusion equation

(5.6) $$\begin{cases} u_t = k u_{xx}, & x \in \mathbb{R},\ t > 0, \\ u(x,0) = \varphi(x). \end{cases}$$

As we have shown in the previous section, from the physical point of view this problem describes diffusion in an infinitely long tube or heat propagation in an infinitely long bar. In the former case, the function $\varphi(x)$ describes the initial concentration of the diffusing substance, in the latter case, it represents the initial distribution of temperature in the bar.

Since the general solution is not known for the diffusion equation, we proceed in a completely different way than we did in the case of the wave equation. We start with solving problem (5.6) with a special "unit step" initial condition $\varphi(x)$. More precisely, we solve the problem

(5.7) $$\begin{cases} w_t = k w_{xx}, & x \in \mathbb{R},\ t > 0, \\ w(x,0) = 0 & \text{for } x < 0; \quad w(x,0) = w_0 \equiv 1 \quad \text{for } x > 0. \end{cases}$$

To derive a solution of this special problem, we use the fact that any physical law can be transferred into a *dimensionless form*. In other words, if we consider an equation linking physical quantities $q_1, \ldots, q_m$ of certain dimensions (time, length, mass,

Section 5.2 Cauchy Problem on the Real Line

etc.), we can find an equivalent relation with dimensionless quantities derived from $q_1, \ldots, q_m$. This process is known as *Buckingham Pi Theorem* (see, e.g., [13]) and we illustrate it by a simple example. Let us imagine an object that was thrown upright at time $t = 0$ at speed v. The height h of the object at time t is given by the formula

$$h = -\frac{1}{2}gt^2 + vt.$$

The constant g represents the gravity acceleration. The quantities used here are h, t, v and g with dimensions of length, time, length per time, and length per time-squared, respectively. This law can be written equivalently also as

$$\frac{h}{vt} = -\frac{1}{2}\left(\frac{gt}{v}\right) + 1.$$

If we denote

$$P_1 := \frac{h}{vt} \quad \text{and} \quad P_2 := \frac{gt}{v},$$

then P_1, P_2 are quantities without dimensions and the original equation has the form

$$P_1 = -\frac{1}{2}P_2 + 1.$$

A similar process can be applied also in the case of our special problem (5.7). The quantities considered are x, t, w, w_0, k, which have—for the heat transfer model—dimensions of length, time, temperature, again temperature, and length-squared per time, respectively. It is clear that w/w_0 is a dimensionless quantity. The only other dimensionless quantity derived from the remaining parameters is $x/\sqrt{4kt}$ (constant 4 is here only for simplification of further relations). We can thus expect the solution of (5.7) to have the form of a combination of these dimensionless variables, that is

$$\frac{w}{w_0} = f\left(\frac{x}{\sqrt{4kt}}\right),$$

where f is for now an unknown function that has to be determined. We recall that $w_0 \equiv 1$. Now, let us introduce a substitution

$$w = f(z), \quad z = \frac{x}{\sqrt{4kt}}$$

and put it into the equation of problem (5.7). According to the chain rule, we find

$$w_t = f'(z)z_t = -\frac{1}{2}\frac{x}{\sqrt{4kt^3}}f'(z),$$

$$w_x = f'(z)z_x = \frac{1}{\sqrt{4kt}}f'(z), \quad w_{xx} = \frac{\partial}{\partial x}w_x = \frac{1}{4kt}f''(z).$$

If we substitute into (5.7) and simplify, we obtain an ODE

$$f''(z) + 2zf'(z) = 0$$

for an unknown function $f(z)$. We easily derive

$$f(z) = c_1 \int_0^z e^{-s^2}\,ds + c_2,$$

where c_1, c_2 are integration constants. (The reader is asked to do it in detail.) Thus we obtain a solution of the Cauchy problem (5.7) in the form

$$w(x,t) = c_1 \int_0^{x/\sqrt{4kt}} e^{-s^2}\,ds + c_2.$$

To determine the constants c_1, c_2, we use the initial condition. Let us consider a fixed negative x and pass to the limit for $t \to 0+$; then

$$0 = w(x,0) = c_1 \int_0^{-\infty} e^{-s^2}\,ds + c_2.$$

Conversely, for a fixed positive x and $t \to 0+$ we have

$$1 = w(x,0) = c_1 \int_0^{\infty} e^{-s^2}\,ds + c_2.$$

Since

$$\int_0^{\infty} e^{-s^2}\,ds = \frac{\sqrt{\pi}}{2},$$

we easily determine $c_1 = 1/\sqrt{\pi}$, $c_2 = 1/2$. Hence we obtain a formula for the solution of problem (5.7):

(5.8) $$w(x,t) = \frac{1}{2} + \frac{1}{\sqrt{\pi}} \int_0^{x/\sqrt{4kt}} e^{-s^2}\,ds.$$

Using the so called *error function*

$$\mathrm{erf}(z) = \frac{2}{\sqrt{\pi}} \int_0^z e^{-s^2}\,ds,$$

solution (5.8) can be written in an equivalent form

(5.9) $$w(x,t) = \frac{1}{2}\left(1 + \mathrm{erf}\left(\frac{x}{\sqrt{4kt}}\right)\right).$$

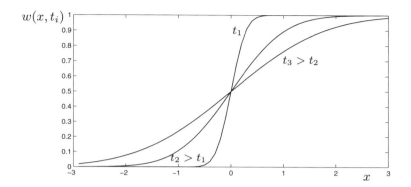

Figure 5.1 *Temperature profile on several time levels for a step initial temperature.*

Several time levels of solution (5.9) are depicted in Figure 5.1.

Now, we come to the second step of derivation of a solution of the general Cauchy problem (5.6). Later (see Chapter 9), we will support our considerations by arguments based on the Fourier transform. For now, however, we put up with an intuitive approach based on physical reasoning. First, let us notice that if w solves the diffusion equation and the partial derivatives w_{xt}, w_{xxx} exist, then also w_x solves the same equation. Namely,

$$0 = (w_t - kw_{xx})_x = (w_x)_t - k(w_x)_{xx}.$$

Thus, the function

$$G(x,t) := w_x(x,t),$$

where $w(x,t)$ is given by formula (5.8), must solve the diffusion equation as well. By direct differentiation with respect to x we obtain

(5.10) $$G(x,t) = \frac{1}{\sqrt{4\pi kt}} e^{-x^2/(4kt)}.$$

The function G is called the *heat (diffusion) kernel* or the *fundamental solution* of the diffusion equation (sometimes we can also meet the terms Green's function, source function, Gaussian, or propagator). Its graph for any fixed $t > 0$ is a "bell-shaped" curve (see Figure 5.2), which has the property that the area below is for each t equal to one:

$$\int_{-\infty}^{\infty} G(x,t)\, dx = 1, \quad t > 0.$$

For $t \to 0+$, $G(x,t)$ "approaches" the so called *Dirac distribution* $\delta(x)$.

Remark 5.1 Let us remark that the Dirac distribution can be understood intuitively as a "generalized function" which achieves an infinite value at point 0, is equal to zero at the other points, and $\int_{-\infty}^{\infty} \delta(x)\,\mathrm{d}x = 1$. (Note that the integral has to be understood in a more general sense than in the case of the Riemann integral!) The problem of correct definition of the Dirac distribution and the word "approaches" is a matter of the theory of distributions and goes beyond the scope of this text.

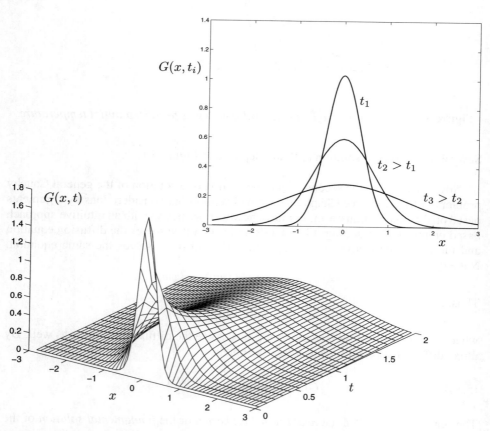

Figure 5.2 *Fundamental solution of the diffusion equation (here, with the choice $k = 0.5$).*

From the physical point of view, the function $G(x,t)$ describes the distribution of temperature as a reaction to the initial unit point source of heat at the point $x = 0$. Further, we observe that the diffusion equation is invariant with respect to translation. Thus, the shifted diffusion kernel $G(x - y, t)$ also solves the diffusion equation and represents a reaction to the initial unit point source of heat at a fixed, but arbitrary, point y. If the initial source is not unit, but has a magnitude $\varphi(y)$, then its contribution at a point x and time t is given by the function $\varphi(y)G(x - y, t)$. The area below the temperature curve is then equal to $\varphi(y)$, where y is the point where the source is

Section 5.2 Cauchy Problem on the Real Line

located.

Let us suppose now that the initial temperature φ in problem (5.6) represents a continuous distribution of heat sources $\varphi(y)$ at points $y \in \mathbb{R}$. Then we obtain the resulting distribution of temperature as a "sum" of all reactions $\varphi(y)G(x-y,t)$ to particular sources $\varphi(y)$ at all points y. That is,

(5.11)
$$u(x,t) = \int_{-\infty}^{\infty} \varphi(y) G(x-y,t)\, dy$$
$$= \frac{1}{\sqrt{4\pi kt}} \int_{-\infty}^{\infty} \varphi(y) e^{-\frac{(x-y)^2}{4kt}} dy.$$

We have just derived intuitively the following basic existence result for the diffusion equation.

Theorem 5.2 *Let φ be a bounded continuous function on $\mathbb{R}$. The Cauchy problem (5.6) for the diffusion equation on the real line has a classical solution given by formula (5.11).*

Remark 5.3 It can be shown that the function $u(x,t)$ given by formula (5.11) solves problem (5.6) also in the case that φ is only *piecewise continuous*. Then, at the points of discontinuity, the solution converges for $t \to 0+$ to the arithmetical average of the left and right limits of the function φ, that is,

$$u(x,t) \to \frac{1}{2}[\varphi(x-) + \varphi(x+)].$$

The decaying character of $G(x-y,t)$ as $|y| \to +\infty$ and x is fixed allows to show that the integral in (5.11) is finite (and expresses the solution of the Cauchy problem (5.6)) also for certain unbounded initial conditions φ.

Concerning the uniqueness, it can be proved that there exists only one *bounded* solution of the Cauchy problem (5.6). In general, without any conditions at infinity, the uniqueness does not hold true (see, e.g., [11]).

Now, let us mention some fundamental properties of the solution of the Cauchy problem for the diffusion equation. Relation (5.11) has an integral form and it can be seen that it cannot be expressed analytically (that is, it cannot be written in terms of elementary functions) for majority of initial conditions. If we are interested in the form of a solution in a particular case, we have to integrate numerically, or to use some program package.

Further, let us notice that, for $t > 0$, the solution $u(x,t)$ is non-zero at an arbitrary point x even if the initial condition φ is non-zero only on a "small interval". It would mean that the *heat propagates at infinite speed*, and also the *diffusion has infinite speed*. But this phenomenon does not correspond to reality and reflects the fact that the diffusion equation is only an approximate model of the real process. On the other hand, for small t, the influence of the initial distribution trails away quickly with growing distance. Thus, we can say the model is precise enough to be applicable from the practical point of view.

Another property of solution (5.11) is its smoothness. Regardless of the smoothness of the function φ, the solution u is of the class C^∞ for $t > 0$ (that is, infinitely times continuously differentiable in both variables).

Example 5.4 We solve the problem

(5.12)
$$\begin{cases} u_t = k u_{xx}, & x \in \mathbb{R},\ t > 0, \\ u(x,0) = \varphi(x) = \begin{cases} 1 & \text{for } |x| < 1, \\ 0 & \text{for } |x| \geq 1. \end{cases} \end{cases}$$

It is a Cauchy problem for the diffusion equation with a piecewise continuous (possibly *non-smooth*) *initial condition*. The solution can be determined by substituting the initial condition into formula (5.11):

$$u(x,t) = \frac{1}{\sqrt{4\pi kt}} \int_{-1}^{1} e^{-\frac{(x-y)^2}{4kt}}\, dy.$$

If we introduce $p = (x-y)/\sqrt{4kt}$, we obtain the expression

$$u(x,t) = \frac{1}{\sqrt{\pi}} \int_{\frac{x-1}{\sqrt{4kt}}}^{\frac{x+1}{\sqrt{4kt}}} e^{-p^2}\, dp$$

or

$$u(x,t) = \frac{1}{2}\left[\operatorname{erf}\left(\frac{x+1}{\sqrt{4kt}}\right) - \operatorname{erf}\left(\frac{x-1}{\sqrt{4kt}}\right)\right].$$

□

Remark 5.5 The graph of the solution from Example 5.4 is sketched in Figure 5.3 (for $k = 2$). Let us mention its basic features. The initial distribution of temperature was a piecewise continuous function, however, it is immediately completely smoothened. After an arbitrarily small time, the solution is non-zero on the whole real line, although the initial condition is non-zero only on the interval $(-1, 1)$. Further, it is evident that the solution achieves its maximal value at time $t = 0$ and, with growing time, it is being "spread".

Example 5.6 (Stavroulakis, Tersian [18]) We solve the problem

(5.13)
$$\begin{cases} u_t = k u_{xx}, & x \in \mathbb{R},\ t > 0, \\ u(x,0) = e^{-x}. \end{cases}$$

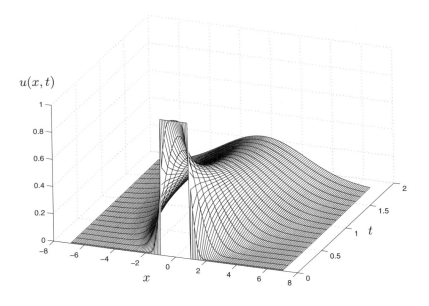

Figure 5.3 *Solution of Example 5.4 with $k = 2$.*

Let us observe that the given *initial condition is not bounded* on $\mathbb{R}$ (cf. Remark 5.3). If we use formula (5.11), we obtain the solution in the form

$$(5.14) \qquad u(x,t) = \frac{1}{\sqrt{4\pi kt}} \int_{-\infty}^{\infty} e^{-\frac{(x-y)^2}{4kt}} e^{-y} \, dy.$$

The integral on the right-hand side can be calculated and we obtain

$$(5.15) \qquad u(x,t) = e^{kt-x}$$

(cf. Exercise 3 in Section 5.4). In this case, the solution does not decrease with growing time, but it propagates in the direction of the positive half-axis x. □

5.3 Diffusion Equation with Sources

We start with a formulation of the basic existence result.

Theorem 5.7 Let $f = f(x,t)$ and $\varphi = \varphi(x)$ be bounded and continuous functions. The Cauchy problem for the nonhomogeneous diffusion equation

(5.16)
$$u_t - k u_{xx} = f(x,t), \quad x \in \mathbb{R}, \; t > 0,$$
$$u(x,0) = \varphi(x)$$

has a classical solution given by the formula

(5.17)
$$u(x,t) = \int_{-\infty}^{\infty} G(x-y,t)\varphi(y)\,dy + \int_0^t \int_{-\infty}^{\infty} G(x-y,t-s) f(y,s)\,dy\,ds,$$

where G is the diffusion kernel.

First, we derive formula (5.17) using the *operator method* (see Section 4.3) based on the analogue with the solution of ODE

(5.18)
$$\frac{dv}{dt} + Av(t) = f(t), \quad v(0) = \varphi,$$

where A is a constant and $\varphi \in \mathbb{R}$. We easily find out that the corresponding solution has the form

(5.19)
$$v(t) = S(t)\varphi + \int_0^t S(t-s) f(s)\,ds,$$

where $S(t) = e^{-tA}$.

Now, we turn back to the original diffusion problem (5.16). The solution of the homogeneous diffusion equation can be written in the form

$$u(x,t) = \int_{-\infty}^{\infty} G(x-y,t)\varphi(y)\,dy.$$

Similarly to Section 4.3, we set

(5.20) $\quad u(x,t) = S(t)\varphi(x), \quad \text{i.e.,} \quad S(t)\varphi(x) = \int_{-\infty}^{\infty} G(x-y,t)\varphi(y)\,dy.$

The operator $S(t)$, called the *source operator*, transforms the function φ into a solution of the homogeneous diffusion equation and hence is an obvious analogue of the function $S(t)$. If we use this analogue, we can expect that the solution of the nonhomogeneous diffusion equation will have the form (in accordance with relation (5.19))

$$u(x,t) = S(t)\varphi(x) + \int_0^t S(t-s) f(x,s)\,ds,$$

Section 5.3 Diffusion Equation with Sources

which is (after substituting for $S(t)$) the derived formula (5.17):

$$u(x,t) = \int_{-\infty}^{\infty} G(x-y,t)\varphi(y)\,dy + \int_0^t \int_{-\infty}^{\infty} G(x-y,t-s)f(y,s)\,dy\,ds.$$

Now, the only point left is to verify that (5.17) really solves problem (5.16). For simplicity, let us assume $\varphi(x) \equiv 0$ and consider only the influence of the right-hand side f. First, we verify that the solution fulfills the equation. We use the rule for differentiation of the integral with respect to a parameter t, thus obtaining

$$\frac{\partial u}{\partial t}(x,t) = \frac{\partial}{\partial t}\int_0^t \int_{-\infty}^{\infty} G(x-y,t-s)f(y,s)\,dy\,ds$$

$$= \int_0^t \int_{-\infty}^{\infty} \frac{\partial G}{\partial t}(x-y,t-s)f(y,s)\,dy\,ds$$

$$+ \lim_{s\to t}\int_{-\infty}^{\infty} G(x-y,t-s)f(y,s)\,dy.$$

The reader should notice that the exchange of differentiation is not always possible, in particular, we have to be careful of the singularities of the function $G(x-y,t-s)$ at the time $t=s$! If, in the first integral, we use the fact that G solves the diffusion equation and substitute $s := t - \epsilon$, we obtain

$$\frac{\partial u}{\partial t}(x,t) = \int_0^t \int_{-\infty}^{\infty} k\frac{\partial^2 G}{\partial x^2}(x-y,t-s)f(y,s)\,dy\,ds$$

$$+ \lim_{\epsilon\to 0}\int_{-\infty}^{\infty} G(x-y,\epsilon)f(y,t-\epsilon)\,dy.$$

Now, we can change the order of integration and differentiation in the first term on the right-hand side. Moreover, it can be rigorously shown (as follows from the theory of distributions, see, e.g., Friedlander, Joshi [10]) that for $\epsilon \to 0$ the function $G(x-y,\epsilon)$ converges to the Dirac distribution at the point x. By continuity of f, $f(y,t-\epsilon)$ converges to $f(y,t)$. Finally, we obtain the relation

$$\frac{\partial u}{\partial t}(x,t) = k\frac{\partial^2}{\partial x^2}\int_0^t \int_{-\infty}^{\infty} G(x-y,t-s)f(y,s)\,dy\,ds + f(x,t)$$

$$= k\frac{\partial^2 u}{\partial x^2} + f(x,t),$$

which is exactly the nonhomogeneous diffusion equation of problem (5.16).

Further, we have to verify the initial condition. Due to the properties of the diffusion kernel G, the first term in (5.17) converges for $t \to 0+$ to the initial condition $\varphi(x)$. The second term is an integral from 0 to 0, thus

$$\lim_{t\to 0+} u(x,t) = \varphi(x) + \int_0^0 \cdots = \varphi(x),$$

which we wanted to prove.

If we substitute the concrete form of diffusion kernel (5.10) into expression (5.17), we obtain the solution of the Cauchy problem for the nonhomogeneous diffusion equation (5.16) in the form

(5.21)
$$u(x,t) = \int_{-\infty}^{\infty} \frac{1}{\sqrt{4\pi kt}} e^{-\frac{(x-y)^2}{4kt}} \varphi(y)\,dy$$
$$+ \int_0^t \int_{-\infty}^{\infty} \frac{1}{\sqrt{4\pi k(t-s)}} e^{-\frac{(x-y)^2}{4k(t-s)}} f(y,s)\,dy\,ds.$$

Remark 5.8 The reader should notice that above we have used formally some assertions like the derivative of the integral with respect to the parameter, passage to the limit under the integral sign, properties of the Dirac distribution, etc., without checking carefully the assumptions. Similarly to the case of Theorem 5.2, the existence result in Theorem 5.7 still holds if φ and f are more general functions.

5.4 Exercises

1. Consider heat conduction in a rod with perfect lateral insulation, no internal heat sources, and specific heat, mass density and thermal conductivity as functions of x, that is, $c(x)$, $\rho(x)$ and $K(x)$. Start with the energy conservation law and derive a new form of the heat equation.

2. Consider a rod with perfect lateral insulation, but with the cross-sectional area dependent on x, that is, $A(x)$. Derive a one-dimensional conservation law for a tube with variable cross-section (cf. Section 1.4), and use Fourier's heat law as a constitutive law to obtain the corresponding heat equation.

3. Show that the solution from Example 5.6 given by (5.14) assumes the simple form of (5.15).

   Here, use $\frac{1}{\sqrt{4\pi kt}} \int_{-\infty}^{\infty} e^{-\frac{(x-y)^2}{4kt}} e^{-y}\,dy = \frac{1}{\sqrt{4\pi kt}} \int_{-\infty}^{\infty} e^{-\frac{(y-x+2kt)^2}{4kt}} e^{kt-x}\,dy$ and the substitution $s = \frac{y-x+2kt}{\sqrt{4kt}}$.

4. Verify that the function
$$u(x,t) = \frac{1}{\sqrt{4\pi kt}} e^{-x^2/4kt}$$
solves the diffusion equation $u_t = k u_{xx}$ on the domain $-\infty < x < \infty$, $t > 0$. Observe how the diffusion parameter k influences the solution.

Section 5.4 Exercises

5. For which values of a and b is the function $u(x,t) = e^{at} \sin bx$ a solution of the diffusion equation $u_t - u_{xx} = 0$?

   $[a + b^2 = 0]$

6. Suppose $|\varphi(x)| \leq M$ for all $x \in \mathbb{R}$ (M is a positive constant). Use the fact that $|\int f| \leq \int |f|$ and show that the solution of the Cauchy problem (5.6) for the diffusion equation satisfies $|u(x,t)| \leq M$ for all $x \in \mathbb{R}$, $t > 0$.

7. Verify that
$$\int_{-\infty}^{\infty} G(x,t)\,dx = 1, \quad t > 0.$$

8. Solve the diffusion equation $u_t = k u_{xx}$ with the initial condition
$$\varphi(x) = 1 \text{ for } x > 0, \quad \varphi(x) = 3 \text{ for } x < 0.$$
   Write the solution using the error function $\text{erf}(x)$.

   $[u(x,t) = 2 - \text{erf}(\frac{x}{\sqrt{4kt}})]$

9. Solve the diffusion equation $u_t = k u_{xx}$ with the initial condition $\varphi(x) = e^{3x}$.

   $[u(x,t) = e^{3x+9kt}]$

10. Solve the diffusion equation $u_t = k u_{xx}$ with the initial condition
$$\varphi(x) = e^{-x} \text{ for } x > 0, \quad \varphi(x) = 0 \text{ for } x < 0.$$

    $[u(x,t) = \frac{1}{2}e^{kt-x}(1 - \text{erf}(\frac{2kt-x}{\sqrt{4kt}}))]$

11. Solve the diffusion equation $u_t = u_{xx}$ with the initial condition
$$\varphi(x) = \begin{cases} 1-x, & 0 \leq x \leq 1, \\ 1+x, & -1 \leq x \leq 0, \\ 0, & |x| \geq 1. \end{cases}$$
    Show that $u(x,t) \to 0$ as $t \to \infty$ for every x.

12. Using the substitution $u(x,t) = e^{-bt} v(x,t)$, solve the diffusion equation
$$u_t - k u_{xx} + bu = 0, \quad u(x,0) = \varphi(x).$$
    Here, b is a positive constant representing dissipation.

13. Using the substitution $y = x - Vt$, solve the heat equation
$$u_t - k u_{xx} + V u_x = 0, \quad u(x,0) = \varphi(x).$$
    Here, V is a positive constant representing convection.

    $[u(x,t) = \frac{1}{\sqrt{4\pi kt}} \int_{-\infty}^{\infty} e^{-(x-Vt-z)^2/(4kt)} \varphi(z)\,dz]$

14. Show that the equation $u_t = k(t)u_{xx}$ can be transformed into a diffusion equation by changing the time variable t into

$$\tau = \int_0^t k(\eta)\,d\eta.$$

Similarly, show that the equation $u_t = ku_{xx} - b(t)u_x$ can be transformed into a diffusion equation by changing the spatial variable x into

$$\xi = x - \int_0^t b(\eta)\,d\eta.$$

15. Find the solution of the problem

$$u_t - ku_{xx} = \sin x, \quad x \in \mathbb{R},\ t > 0, \quad u(x,0) = 0.$$

$[u(x,t) = \tfrac{1}{k}(1 - e^{-kt})\sin x]$

16. Show that the transport equation with diffusion and decay

$$u_t = ku_{xx} - cu_x - \lambda u$$

can be transformed into a diffusion equation by a substitution

$$u(x,t) = w(x,t)e^{\alpha x - \beta t}$$

with $\alpha = \tfrac{c}{2k}$ and $\beta = \lambda + \tfrac{c^2}{4k}$.

Exercises 1, 2 are taken from Keane [12], 5, 7, 10, 14–16 from Logan [14], and 8–10, 12, 13 from Strauss [19].

6 Laplace and Poisson Equations in Two Dimensions

In the previous chapters we have met the basic representatives of hyperbolic equations (the wave equation) and of parabolic equations (the diffusion equation). This chapter is devoted to the simplest elliptic equation in two dimensions, that is, the Laplace equation.

6.1 Steady States and Laplace and Poisson Equations

Studying dynamical models, we are often interested only in the behavior in the so called *steady (stationary, or equilibrium) state*, that is, in the state when the *solution does not depend on time* ($u_t = u_{tt} = 0$). In such a case, the (in general, multidimensional) diffusion equation $u_t = k\Delta u$ as well as the wave equation $u_{tt} = c^2 \Delta u$ are reduced to the *Laplace equation*

$$\boxed{\Delta u = 0,}$$

which in two dimensions reduces to $u_{xx} + u_{yy} = 0$. Solutions of the Laplace equation are the so called *harmonic functions*.

Let us consider a plane body that is heated in an oven. We assume that the temperature in the oven is not the same everywhere (it is not spatially constant). After a certain time, the temperature in the body achieves the steady state which will be described by a harmonic function $u(x, y)$. In the case that the temperature in the oven is spatially constant, the steady state corresponds also to $u(x, y) = \text{const}$. In one-dimensional case, we can imagine a laterally isolated rod in which the heat interchange with the neighborhood acts only at its ends. The function u describing the temperature in the rod then depends only on x. The Laplace equation has thus the form

$$u_{xx} = 0$$

and its solution is any linear function $u(x) = c_1 x + c_2$. In higher dimensions, the situation is much more interesting.

The steady state can be studied also in the case when the model includes time-independent sources. The nonhomogeneous analogue of the Laplace equation with a given function f is the *Poisson equation*

$$\boxed{\Delta u = f.}$$

The Laplace and Poisson equations appear, for instance, in the following models.

Electrostatics. The Maxwell equations

$$\text{rot } \boldsymbol{E} = 0, \quad \text{div } \boldsymbol{E} = \frac{\rho}{\varepsilon}$$

describe an electrostatic vector field of intensity $\boldsymbol{E}$ in a medium of constant permittivity ε; ρ represents the volume density of the electric charge. The first equation implies the existence of the so called *electric potential*, which is a scalar function ϕ satisfying the relation $\boldsymbol{E} = -\text{grad } \phi$. If we substitute ϕ into the latter equation, we obtain

$$\Delta \phi = \text{div}(\text{grad } \phi) = -\text{div } \boldsymbol{E} = -\frac{\rho}{\varepsilon},$$

which is the Poisson equation with the right-hand side $f = -\frac{\rho}{\varepsilon}$.

Steady Flow. Let us assume that we model an irrotational flow described by the equation rot $\boldsymbol{v} = 0$, where $\boldsymbol{v}$ is the flow speed (independent of time). This equation implies the existence of a scalar function ϕ (the so called *velocity potential*) satisfying $\boldsymbol{v} = -\text{grad } \phi$. Let, moreover, the flowing liquid be incompressible (for example, water) and let the flow be solenoidal (without sources and sinks). Then div $\boldsymbol{v} = 0$. If we substitute here the potential ϕ, we can write $\Delta \phi = -\text{div}(\text{grad } \phi) = -\text{div } \boldsymbol{v} = 0$. Thus, we obtain the Laplace equation $\Delta \phi = 0$.

Harmonic Components of a Holomorphic Function of One Complex Variable. Let us write $z = x + iy$ and

$$f(z) = u(z) + iv(z) = u(x+iy) + iv(x+iy),$$

where u and v are real functions of a complex variable z. Since the Gauss complex plane can be identified with $\mathbb{R}^2$, we view $u(z) = u(x,y)$, $v(z) = v(x,y)$ as functions of two independent real variables x and y. Theory of functions of a complex variable says that a holomorphic function f on a domain Ω (it means a complex function f that has a derivative $f'(z)$ for every $z \in \Omega$) can be expanded locally into a power series with a center $z_0 \in \Omega$. If $z_0 = 0$, then this expansion of f takes the form

$$f(z) = \sum_{n=0}^{\infty} a_n z^n,$$

where a_n are complex constants. If we substitute for f and z, we obtain

$$u(x,y) + iv(x,y) = \sum_{n=0}^{\infty} a_n (x+iy)^n.$$

Differentiation of this series (the reader is asked to verify it) leads to

$$\frac{\partial u}{\partial x} = \frac{\partial v}{\partial y} \quad \text{and} \quad \frac{\partial u}{\partial y} = -\frac{\partial v}{\partial x},$$

which are the so-called *Cauchy-Riemann* conditions of differentiability of a complex function of a complex variable. By further differentiation we find out that

$$u_{xx} = v_{yx} = v_{xy} = -u_{yy},$$

and thus $\Delta u = 0$. Similarly, $\Delta v = 0$. These formal calculations illustrate that both the real and imaginary parts of a holomorphic function are harmonic functions.

6.2 Invariance of the Laplace Operator, Its Transformation into Polar Coordinates

The Laplace operator Δ (called also the *Laplacian*) is invariant with respect to transformations consisting of translations and rotations. The *translation* in plane by a vector (a, b) is given by the transformation

$$x' = x + a, \qquad y' = y + b.$$

Obviously, $u_{xx} + u_{yy} = u_{x'x'} + u_{y'y'}$.

The *rotation* in plane through an angle α is given by the transformation

$$\begin{aligned} x' &= x \cos \alpha + y \sin \alpha, \\ y' &= -x \sin \alpha + y \cos \alpha. \end{aligned}$$

Using the chain rule, we derive

$$\begin{aligned} u_x &= u_{x'} \cos \alpha - u_{y'} \sin \alpha, \\ u_y &= u_{x'} \sin \alpha + u_{y'} \cos \alpha, \\ u_{xx} &= u_{x'x'} \cos^2 \alpha - 2 u_{x'y'} \sin \alpha \cos \alpha + u_{y'y'} \sin^2 \alpha, \\ u_{yy} &= u_{x'x'} \sin^2 \alpha + 2 u_{x'y'} \sin \alpha \cos \alpha + u_{y'y'} \cos^2 \alpha, \end{aligned}$$

and, summing up, we obtain

$$u_{xx} + u_{yy} = u_{x'x'} + u_{y'y'}.$$

For these properties, the Laplace operator is used in modeling of *isotropic* physical phenomena.

The invariancy with respect to rotation suggests that the Laplace operator could assume a simple form in *polar coordinates*, in particular, in the radially symmetric case. The transformation formulas between the Cartesian and polar coordinates have the form

$$x = r \cos \theta, \qquad y = r \sin \theta,$$

and the corresponding Jacobi matrix J and its inverse J^{-1} are

$$J = \begin{pmatrix} x_r & y_r \\ x_\theta & y_\theta \end{pmatrix} = \begin{pmatrix} \cos \theta & \sin \theta \\ -r \sin \theta & r \cos \theta \end{pmatrix},$$

$$J^{-1} = \begin{pmatrix} r_x & \theta_x \\ r_y & \theta_y \end{pmatrix} = \begin{pmatrix} \cos \theta & -\dfrac{\sin \theta}{r} \\ \sin \theta & \dfrac{\cos \theta}{r} \end{pmatrix}.$$

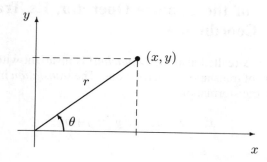

Figure 6.1 *Polar coordinates r and θ.*

Written formally in symbols, we easily find out by differentiation that

$$\frac{\partial^2}{\partial x^2} = \left(\cos\theta\frac{\partial}{\partial r} - \frac{\sin\theta}{r}\frac{\partial}{\partial\theta}\right)^2$$

$$= \cos^2\theta\frac{\partial^2}{\partial r^2} - 2\frac{\sin\theta\cos\theta}{r}\frac{\partial^2}{\partial r\partial\theta} + \frac{\sin^2\theta}{r^2}\frac{\partial^2}{\partial\theta^2} + 2\frac{\sin\theta\cos\theta}{r^2}\frac{\partial}{\partial\theta} + \frac{\sin^2\theta}{r}\frac{\partial}{\partial r},$$

$$\frac{\partial^2}{\partial y^2} = \left(\sin\theta\frac{\partial}{\partial r} + \frac{\cos\theta}{r}\frac{\partial}{\partial\theta}\right)^2$$

$$= \sin^2\theta\frac{\partial^2}{\partial r^2} + 2\frac{\sin\theta\cos\theta}{r}\frac{\partial^2}{\partial r\partial\theta} + \frac{\cos^2\theta}{r^2}\frac{\partial^2}{\partial\theta^2} - 2\frac{\sin\theta\cos\theta}{r^2}\frac{\partial}{\partial\theta} + \frac{\cos^2\theta}{r}\frac{\partial}{\partial r}.$$

Summing these operators, we obtain

(6.1) $$\boxed{\Delta = \frac{\partial^2}{\partial x^2} + \frac{\partial^2}{\partial y^2} = \frac{\partial^2}{\partial r^2} + \frac{1}{r}\frac{\partial}{\partial r} + \frac{1}{r^2}\frac{\partial^2}{\partial\theta^2}.}$$

6.3 Solution of Laplace and Poisson Equations in $\mathbb{R}^2$

Laplace Equation. Using the similarity to the wave equation (4.8), we can find a "general solution" of the Laplace equation in the xy-plane. Indeed, the wave and Laplace equations are formally identical provided we set the speed of wave propagation c to be equal to the imaginary unit $i = \sqrt{-1}$. Thus, according to (4.13), we can conclude that any function of the form

$$u(x,y) = f(x+iy) + g(x-iy)$$

Section 6.3 Solution of Laplace and Poisson Equations in $\mathbb{R}^2$

solves the Laplace equation $\Delta u = 0$ in two dimensions. Here f and g are arbitrary differentiable functions of a complex variable. Since $x - iy$ is the complex conjugate number to $x + iy$, the general solution of the Laplace equation can be written simply as

$$\boxed{u(x, y) = f(x + iy).}$$

However, further analysis is a subject of the theory of complex functions and exceeds the scope of this text.

In the radially symmetric case (when the functions considered do not depend on the angle θ), the Laplace equation in polar coordinates reduces to

$$u_{rr} + \frac{1}{r} u_r = 0.$$

Multiplying by $r > 0$, we obtain the equation

$$r u_{rr} + u_r = 0$$

which is equivalent to

$$(r u_r)_r = 0.$$

This is an ODE that is easy to solve by direct integration:

$$r u_r = c_1$$

and

$$\boxed{u(r) = c_1 \ln r + c_2.}$$

Thus, the radially symmetric harmonic functions in $\mathbb{R}^2 \setminus \{0\}$ are the constant ones and the logarithm, which will play an important role also in the sequel.

Poisson Equation. The same approaches can be applied also to the *Poisson equation* $\Delta u = f(x, y)$. Considering again the radially symmetric case (when $f = f(\sqrt{x^2 + y^2})$), the problem reduces to the equation

$$u_{rr} + \frac{1}{r} u_r = f(r),$$

which is equivalent to

$$r u_{rr} + u_r = r f(r) \quad \text{or} \quad (r u_r)_r = r f(r).$$

Again, by direct integration, we obtain

$$r u_r = c_1 + \int_0^r s f(s) \, ds$$

and, finally,

$$\boxed{u(r) = c_1 \ln r + c_2 + \int_0^r \frac{1}{\sigma} \int_0^\sigma s f(s) \, ds \, d\sigma.}$$

6.4 Exercises

1. Verify that the given functions solve the two-dimensional Laplace equation.

   (a) $u = x + y$.
   (b) $u = x^2 - y^2$.
   (c) $u = \frac{x}{x^2+y^2}$.
   (d) $u = \ln(x^2 + y^2)$.
   (e) $u = \ln\sqrt{x^2 + y^2}$.
   (f) $u = e^y \cos x$.

2. Decide whether the following functions satisfy the Laplace equation.

   (a) $u = \frac{y}{x^2+y^2}$.
   (b) $u = \frac{1}{\sqrt{x^2+y^2}}$.
   (c) $u = \tan^{-1}(\frac{y}{x})$.
   (d) $u = \tan^{-1}(\frac{y}{x})\frac{y}{x^2+y^2}$.
   (e) $u = xy$.

   [yes for a,c,e]

3. Show that if u and v are harmonic and α and β are numbers, then $\alpha u + \beta v$ is harmonic.

4. Give an example of two harmonic functions u and v such that uv is not harmonic.

5. Show that if u and u^2 are both harmonic, then u must be constant.

6. Show that if u, v and $u^2 + v^2$ are harmonic, then u and v must be constant.

7. If $u(x,y)$ is a solution of the Laplace equation, prove that any partial derivative of $u(x,y)$ with respect to one or more Cartesian coordinates (for example, u_x, u_{xx}, u_{xy}) is also a solution.

8. Consider the problem
$$\begin{cases} u_{xx} + u_{yy} = 0 & \text{in } \mathbb{R} \times (0, \infty), \\ u(x,0) = 0, \quad u_y(x,0) = \frac{\cos nx}{n^2}. \end{cases}$$

   Show that $u_n(x,y) = \frac{1}{n^3} \sinh ny \cos nx$ is the solution, but

   $$\lim_{n \to \infty} \max_{(x,y) \in \mathbb{R} \times [0,\infty)} |u_n(x,y)| = 0$$

   does not hold.

Section 6.4 Exercises

9. Functions z^2, z^3, e^z, $\ln z$ of a complex variable $z = x + iy$ are analytic. Rewrite them in the following way:

$$\begin{aligned} z^2 &= (x^2 - y^2) + i(2xy), \\ z^3 &= (x^3 - 3xy^2) + i(3x^2 y - y^3), \\ e^z &= (e^x \cos y) + i(e^x \sin y), \\ \ln z &= (\ln \sqrt{x^2 + y^2}) + i(\arg z) = \ln r + i\theta, \end{aligned}$$

and show that all of them (as functions of x and y) satisfy the Laplace equation.

10. Show that the function $u(x,y) = \arctan(y/x)$ satisfies the Laplace equation $u_{xx} + u_{yy} = 0$ for $y > 0$. Using this fact, try to find a solution of the Laplace equation on the domain $y > 0$, that satisfies boundary conditions $u(x,0) = 1$ for $x > 0$ and $u(x,0) = -1$ for $x < 0$.

11. Show that $e^{-\xi y}\sin(\xi x)$, $x \in \mathbb{R}$, $y > 0$ is a solution of the Laplace equation for an arbitrary value of the parameter ξ. Prove that the function

$$u(x,y) = \int_0^\infty c(\xi) e^{-\xi y} \sin(\xi x)\, d\xi$$

solves the same equation for an arbitrary function $c(\xi)$ that is bounded and continuous on $[0, \infty)$. (Assumptions on the function c allow to differentiate under the integral.)

12. Solve the equation $u_{xx} + u_{yy} = 1$ in $r < a$ with $u(x,y)$ vanishing on $r = a$.

$$[u(r) = \tfrac{1}{4}(r^2 - a^2)]$$

13. Solve the equation $u_{xx} + u_{yy} = 1$ in the annulus $0 < a < r < b$ with $u(x,y)$ vanishing on both parts of the boundary $r = a$ and $r = b$.

$$[u(r) = \tfrac{r^2}{4} + \tfrac{b^2 - a^2}{4 \ln \frac{a}{b}} \ln r - \tfrac{b^2 \ln a - a^2 \ln b}{4 \ln \frac{a}{b}}]$$

14. (a) Show that if $v(x,y)$ is a harmonic function, then also $u(x,y) = v(x^2 - y^2, 2xy)$ is a harmonic function.

    (b) Using transformation into polar coordinates, show that the transformation $(x,y) \mapsto (x^2 - y^2, 2xy)$ maps the first quadrant onto the half-plane $\{y > 0\}$.

Exercises 1, 6 are taken from Asmar [3], 7 from Keane [12], 8 from Stavroulakis and Tersian [18], 9 from Snider [17], and 12, 13 from Strauss [19].

7 Solutions of Initial Boundary Value Problems for Evolution Equations

7.1 Initial Boundary Value Problems on Half-Line

Let us start with the solution of the diffusion and wave equations on the whole real line. Notice that if the initial condition for the diffusion equation is an even or odd function, then also the solution of the Cauchy problem is an even or odd function, respectively. The same holds also in the case of the Cauchy problem for the wave equation. (The reader is asked to prove both cases.) We will use this observation in solving the initial boundary value problems for the diffusion and wave equations on the half-line with homogeneous boundary conditions.

Diffusion and Heat Flow on the Half-Line. First, let us consider the initial boundary value problem for the heat equation

(7.1)
$$\begin{cases} u_t = ku_{xx}, & x > 0,\ t > 0, \\ u(0,t) = 0, \\ u(x,0) = \varphi(x), \end{cases}$$

which describes the temperature distribution in the half-infinite bar with heat insulated lateral surface. The Dirichlet boundary condition corresponds to the fact that the end $x = 0$ is kept at the zero temperature. We will solve this problem using the so called *reflection method*, which is based on the idea of extending the problem to the whole real line in such a way that the boundary condition $u(0,t) = 0$ is fulfilled by itself. In our case, that is, in the case of the homogeneous Dirichlet boundary condition (see Section 2.1), this means to use the *odd extension* of the initial condition $\varphi(x)$. We define

(7.2) $\qquad \tilde{\varphi}(x) = \begin{cases} \varphi(x), & x > 0, \\ -\varphi(-x), & x < 0, \end{cases} \qquad \tilde{\varphi}(0) = 0.$

Since an odd initial condition corresponds to an odd solution, we obtain $u(0,t) = 0$ automatically for all $t > 0$ (see Figure 7.1). Let us consider the extended problem

(7.3)
$$\begin{cases} v_t = kv_{xx}, & x \in \mathbb{R},\ t > 0, \\ v(x,0) = \tilde{\varphi}(x), \end{cases}$$

the solution of which can be written in the form

$$v(x,t) = \int_{-\infty}^{\infty} G(x-y, t)\tilde{\varphi}(y)\,dy.$$

Section 7.1 Initial Boundary Value Problems on Half-Line

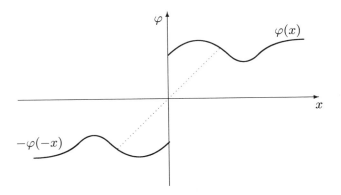

Figure 7.1 *The odd extension.*

If we split the integral into two parts ($y > 0$ and $y < 0$), we obtain

$$v(x,t) = \int_{-\infty}^{0} G(x-y,t)\tilde{\varphi}(y)\,dy + \int_{0}^{\infty} G(x-y,t)\tilde{\varphi}(y)\,dy$$

$$= -\int_{-\infty}^{0} G(x-y,t)\varphi(-y)\,dy + \int_{0}^{\infty} G(x-y,t)\varphi(y)\,dy$$

$$= -\int_{0}^{\infty} G(x+y,t)\varphi(y)\,dy + \int_{0}^{\infty} G(x-y,t)\varphi(y)\,dy$$

$$= \int_{0}^{\infty} [G(x-y,t) - G(x+y,t)]\varphi(y)\,dy.$$

The solution u of the original problem (7.1) is then the restriction of the function v to $x > 0$, i.e.

(7.4) $\quad\boxed{u(x,t) = \int_{0}^{\infty} [G(x-y,t) - G(x+y,t)]\varphi(y)\,dy, \qquad x > 0,\ t > 0.}$

Example 7.1 Figure 7.2 illustrates the function

$$u(x,t) = \text{erf}\left(\frac{x}{\sqrt{4kt}}\right),$$

with $k = 1$, which solves the initial value problem

(7.5) $$\begin{cases} u_t = k u_{xx}, & x > 0,\ t > 0, \\ u(x, 0) = 1, & u(0, t) = 0. \end{cases}$$

□

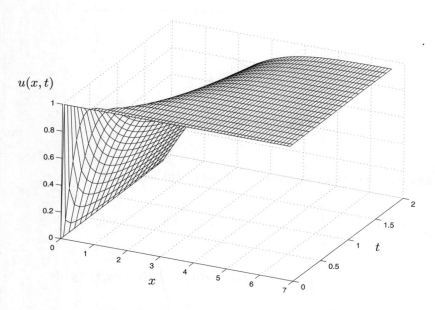

Figure 7.2 *Solution of problem (7.5) with $k = 1$.*

Wave on the Half-Line. The wave equation on the half-line can be solved in the same way. Let us consider a half-infinite string ($x > 0$) whose end $x = 0$ is fixed. The corresponding Cauchy problem takes the form

(7.6) $$\begin{cases} u_{tt} = c^2 u_{xx}, & x > 0,\ t > 0, \\ u(0, t) = 0, \\ u(x, 0) = \varphi(x), & u_t(x, 0) = \psi(x). \end{cases}$$

We again use the method of the *odd extension* of both initial conditions to the whole real line. We introduce

$$\tilde{\varphi}(x) = \begin{cases} \varphi(x), & x > 0, \\ -\varphi(-x), & x < 0, \end{cases} \qquad \tilde{\psi}(x) = \begin{cases} \psi(x), & x > 0, \\ -\psi(-x), & x < 0, \end{cases} \qquad \begin{array}{l} \tilde{\varphi}(0) = 0, \\ \tilde{\psi}(0) = 0, \end{array}$$

and consider the extended problem

(7.7) $$\begin{cases} v_{tt} = c^2 v_{xx}, & x \in \mathbb{R},\ t > 0, \\ v(x, 0) = \tilde{\varphi}(x), & v_t(x, 0) = \tilde{\psi}(x). \end{cases}$$

Section 7.1 Initial Boundary Value Problems on Half-Line

Its solution is given by d'Alembert's formula

$$v(x,t) = \frac{1}{2}[\tilde{\varphi}(x+ct) + \tilde{\varphi}(x-ct)] + \frac{1}{2c}\int_{x-ct}^{x+ct} \tilde{\psi}(y)\,dy.$$

The solution u of the original problem (7.6) will be obtained as the restriction of the function v to $x > 0$.

Let us consider first the region $x > ct$. In the case of points (x,t) from this area, the whole bases of their domains of influence lie in the interval $(0, \infty)$, where $\tilde{\varphi}(x) = \varphi(x)$, $\tilde{\psi}(x) = \psi(x)$. Thus, the solution is here given by the "usual" relation

(7.8) $$u(x,t) = \frac{1}{2}[\varphi(x+ct) + \varphi(x-ct)] + \frac{1}{2c}\int_{x-ct}^{x+ct} \psi(y)\,dy, \quad x > ct.$$

However, in the region $0 < x < ct$ we have $\tilde{\varphi}(x-ct) = -\varphi(ct-x)$ and thus

$$u(x,t) = \frac{1}{2}[\varphi(x+ct) - \varphi(ct-x)] + \frac{1}{2c}\int_{0}^{x+ct} \psi(y)\,dy + \frac{1}{2c}\int_{x-ct}^{0} [-\psi(-y)]\,dy.$$

If we pass in the last integral from the variable y to $-y$, we obtain the solution in the form

(7.9) $$u(x,t) = \frac{1}{2}[\varphi(ct+x) - \varphi(ct-x)] + \frac{1}{2c}\int_{ct-x}^{ct+x} \psi(y)\,dy, \quad 0 < x < ct.$$

The complete solution of problem (7.6) is given by the couple of formulas (7.8), (7.9).

This result can be interpreted in the following graphical way. In the xt plane, we draw the backward characteristics from the point (x,t). If (x,t) lies in the region $x < ct$, one of the characteristics crosses the t axis earlier than it touches the x axis. Relation (7.9) says that there occurs a reflection at the end $x = 0$ and the solution depends on the values of the function φ at the points $ct \pm x$ and on the values of the function ψ on the short interval between these points. Other values of the function ψ have no influence on the solution at the point (x,t) (see Figure 7.3).

In the case of problems with the homogeneous Neumann boundary condition (see Section 2.1), we would proceed analogously. We would use, however, the method of *even extension*, which uses the fact that even initial conditions correspond to even solutions. The latter then fulfil the condition $u_x(0,t) = 0$ at the point $x = 0$ automatically. The derivation of the corresponding solutions is left to the reader.

Example 7.2 Let us consider the initial boundary value problem

(7.10) $$\begin{cases} u_{tt} = c^2 u_{xx}, & x > 0, \ t > 0, \\ u(0,t) = 0, \\ u(x,0) = e^{-(x-3)^2}, & u_t(x,0) = 0. \end{cases}$$

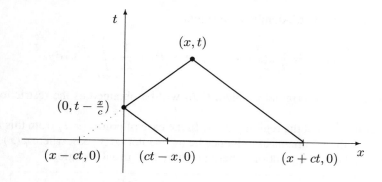

Figure 7.3 *Reflection method for the wave equation.*

Its solution is sketched in Figure 7.4 (here, $c = 2$). Notice the reflection of the initial wave on the boundary line $x = 0$.

□

Example 7.3 Let us consider the initial boundary value problem

(7.11) $$\begin{cases} u_{tt} = c^2 u_{xx}, & x > 0, \ t > 0, \\ u(0,t) = 0, \\ u(x,0) = 0, & u_t(x,0) = \sin x. \end{cases}$$

By direct substitution into formulas (7.8) and (7.9) we find that $u(x,t) = \frac{1}{c}\sin x \sin ct$ in both regions $0 < x < ct$ and $x > ct$. The graph of the solution is shown in Figure 7.5 for the choice $c = 4$. (We have met the same problem on the real line in Chapter 4, see Example 4.2 and Figure 4.2. We recall that the solution was, for its properties, called the *standing wave*.)

□

Problems with Nonhomogeneous Boundary Condition. Let us consider an initial boundary value problem for the diffusion equation on the half-line with a *nonhomogeneous* boundary condition

(7.12) $$\begin{cases} u_t - k u_{xx} = 0, & x > 0, \ t > 0, \\ u(0,t) = h(t), \\ u(x,0) = \varphi(x). \end{cases}$$

In this case we introduce a new function

$$v(x,t) = u(x,t) - h(t),$$

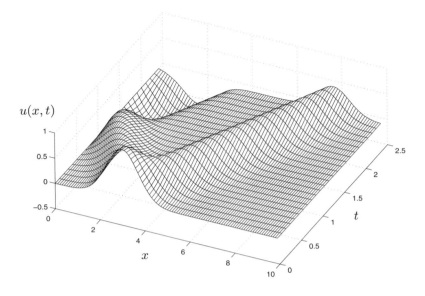

Figure 7.4 *Solution of problem (7.10) for $c = 2$.*

which then solves the problem

$$\begin{cases} v_t - kv_{xx} = -h'(t), & x > 0,\ t > 0, \\ v(0, t) = 0, \\ v(x, 0) = \varphi(x) - h(0). \end{cases}$$

We have thus transferred the influence of the initial condition to the right-hand side of the equation. This means that we solve a *nonhomogeneous* diffusion equation with a *homogeneous* boundary condition. To find a solution of this transformed problem, we can use again the method of odd extension. If the compatibility condition ($\varphi(0) = h(0)$) is not satisfied, we obtain a "generalized" solution which is not continuous at the point $(0, 0)$ (however, it is continuous everywhere else).

In the case of a nonhomogeneous Neumann boundary condition $u_x(0, t) = h(t)$, we would use the transformation $v(x, t) = u(x, t) - xh(t)$, which results in the nonhomogeneous right-hand side $-xh'(t)$, but it ensures the homogeneity of the boundary condition $v_x(0, t) = 0$. The transformed problem is then solved by the method of even extension. The reader is invited to go through the details.

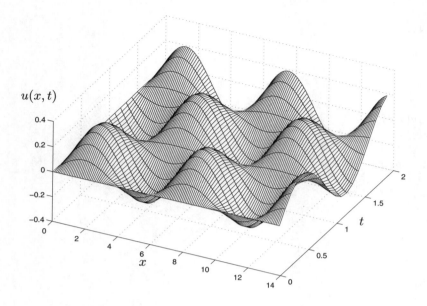

Figure 7.5 *Solution of problem (7.11) for $c = 4$.*

7.2 Initial Boundary Value Problem on Finite Interval, Fourier Method

In this section, we deal with initial boundary value problems for the wave and diffusion equations on finite intervals. Boundary conditions are now given on both ends of the interval considered. Solving these problems can be approached in various ways. One—apparent—way is to apply the *reflection method* (which was discussed in the previous section) on both ends of the interval. In the case of homogeneous Dirichlet boundary conditions, it means to use the odd extension of the initial conditions with respect to both the end-points and, further, to extend all functions periodically to the whole real line. That is, instead of the original initial condition $\varphi = \varphi(x)$, $x \in (0, l)$, we consider the extended function

$$\tilde{\varphi}(x) = \begin{cases} \varphi(x), & x \in (0, l), \\ -\varphi(2l - x), & x \in (l, 2l), \\ -\varphi(-x), & x \in (-l, 0), \\ \varphi(2l + x), & x \in (-2l, -l), \\ 4l - \text{periodic} & \text{elsewhere.} \end{cases}$$

(In the case of Neumann boundary conditions we use the even extension.) Then we can use the formulas for the solution of the particular problems on the real line. However, after substituting back for the "tilde" initial conditions, the resulting formulas become very complicated!

Section 7.2 Initial Boundary Value Problem on Finite Interval, Fourier Method

For example, considering the problem for the *wave equation*, we can obtain the value of the solution at a point (x,t) in the following way: We draw the backwards characteristics from the point (x,t) and reflect them whenever they hit the lines $x = 0$, $x = l$, until we reach the zero time level $t = 0$ (see Figure 7.6(a)). Thus we obtain a couple of points x_1 and x_2. The solution is then determined by the initial displacements at these points, by the initial velocity in the interval (x_1, x_2), and also by the number of reflections. In general, the characteristic lines divide the space-time domain $(0, l) \times (0, \infty)$ into "diamond-shaped" areas, and the solution is given by different formulas on each of these areas (see Figure 7.6(b)).

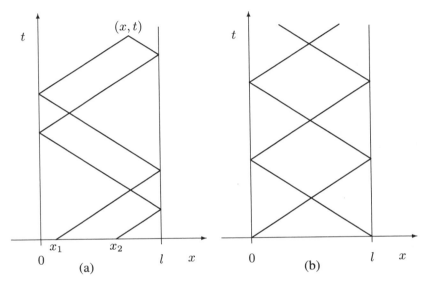

Figure 7.6 *Reflection method for the wave equation on finite interval.*

In the sequel, we focus on the explanation of another standard method, the so called *Fourier method* (sometimes called also the *method of separation of variables*). Its application in solving the initial boundary value problems for the wave equation led to the systematic investigation of trigonometrical series (much later called Fourier series).

Dirichlet Boundary Conditions, Wave Equation. First, we will consider the initial boundary value problem that describes vibrations of a string of finite length l, whose end points are fixed in the "zero position", and whose initial displacement and initial velocity are given by functions $\varphi(x)$ and $\psi(x)$, respectively:

(7.13)
$$\begin{aligned} u_{tt} &= c^2 u_{xx}, \quad 0 < x < l,\ t > 0, \\ u(0,t) &= u(l,t) = 0, \\ u(x,0) &= \varphi(x), \quad u_t(x,0) = \psi(x). \end{aligned}$$

We start with the assumption that there exists a solution of the wave equation in the form

$$u(x,t) = X(x)T(t),$$

where $X = X(x)$ and $T = T(t)$ are real functions of one real variable, their second derivatives exist and are continuous. Variables x and t are thus separated from each other. If we now insert this solution back into the equation, we obtain

$$XT'' = c^2 X'' T$$

and, after dividing by the term $-c^2 XT$ (under the assumption $XT \neq 0$),

$$-\frac{T''(t)}{c^2 T(t)} = -\frac{X''(x)}{X(x)}.$$

This relation says that the function $-\frac{T''}{c^2 T}$, which depends only on the variable t, is equal to the function $-\frac{X''}{X}$, which depends only on the spatial variable x. This equality must hold for all $t > 0$ and $x \in (0, l)$, and thus

$$-\frac{T''}{c^2 T} = -\frac{X''}{X} = \lambda,$$

where λ is a *constant*. The original PDE is thus transformed into the couple of *separated* ODEs for unknown functions $X(x)$ and $T(t)$:

(7.14) $\qquad\qquad\qquad X''(x) + \lambda X(x) = 0,$

(7.15) $\qquad\qquad\qquad T''(t) + c^2 \lambda T(t) = 0.$

Further, we are given homogeneous Dirichlet boundary conditions

(7.16) $\qquad\qquad\qquad X(0) = X(l) = 0,$

since the setting of problem (7.13) implies $u(0,t) = u(l,t) = 0$ for all $t > 0$. First, we will solve the boundary value problem (7.14), (7.16). Since we are evidently not interested in the trivial solution $X(x) \equiv 0$, we exclude the case $\lambda \leq 0$. If $\lambda > 0$, equation (7.14) yields the solution in the form

$$X(x) = C \cos \sqrt{\lambda} x + D \sin \sqrt{\lambda} x$$

and the boundary conditions (7.16) imply

$$X(0) = C = 0 \quad \text{and} \quad X(l) = D \sin \sqrt{\lambda} l = 0.$$

The nontrivial solution can be obtained only in the case

$$\sin \sqrt{\lambda} l = 0,$$

that is,

(7.17) $\qquad\qquad\qquad \lambda_n = \left(\frac{n\pi}{l}\right)^2, \qquad n \in \mathbb{N}.$

Section 7.2 Initial Boundary Value Problem on Finite Interval, Fourier Method

Every such λ_n then corresponds to a solution

(7.18) $$X_n(x) = C_n \sin \frac{n\pi x}{l}, \qquad n \in \mathbb{N},$$

where C_n are arbitrary constants. Now, we go back to equation (7.15). Its solution assumes for $\lambda = \lambda_n$ the form

(7.19) $$T_n(t) = A_n \cos \frac{n\pi ct}{l} + B_n \sin \frac{n\pi ct}{l}, \qquad n \in \mathbb{N},$$

where A_n and B_n are again arbitrary constants. The original PDE of problem (7.13) is then solved by a sequence of functions

$$u_n(x,t) = \left(A_n \cos \frac{n\pi ct}{l} + B_n \sin \frac{n\pi ct}{l} \right) \sin \frac{n\pi x}{l}, \qquad n \in \mathbb{N},$$

which already satisfy the prescribed homogeneous boundary conditions. Let us notice that, instead of $A_n C_n$ and $B_n C_n$, we write only A_n and B_n, since all these real constants are arbitrary. Since the problem is linear, an *arbitrary finite sum* of the form

(7.20) $$u(x,t) = \sum_{n=1}^{N} \left(A_n \cos \frac{n\pi ct}{l} + B_n \sin \frac{n\pi ct}{l} \right) \sin \frac{n\pi x}{l}$$

is also a solution. The function given by formula (7.20) satisfies also the initial conditions provided

(7.21) $$\varphi(x) = \sum_{n=1}^{N} A_n \sin \frac{n\pi x}{l},$$

(7.22) $$\psi(x) = \sum_{n=1}^{N} \frac{n\pi c}{l} B_n \sin \frac{n\pi x}{l}.$$

For arbitrary initial data in this form, problem (7.13) is uniquely solvable and the corresponding solution is given by relation (7.20).

Obviously, conditions (7.21), (7.22) are very restrictive and their satisfaction is hard to ensure. For this reason, we look for a solution of problem (7.13) in the form of an *infinite sum* and we express it in the form of a *Fourier series*

(7.23) $$\boxed{u(x,t) = \sum_{n=1}^{\infty} \left(A_n \cos \frac{n\pi ct}{l} + B_n \sin \frac{n\pi ct}{l} \right) \sin \frac{n\pi x}{l}.}$$

The constants A_n and B_n (or, more precisely, $\frac{n\pi c}{l} B_n$) are then given as the Fourier coefficients of sine expansions of the functions $\varphi(x)$, $\psi(x)$, thus

$$\varphi(x) = \sum_{n=1}^{\infty} A_n \sin \frac{n\pi x}{l},$$

$$\psi(x) = \sum_{n=1}^{\infty} \frac{n\pi c}{l} B_n \sin \frac{n\pi x}{l}.$$

In other words, the solution of the initial boundary value problem for the wave equation can be expressed at every time t in the form of a Fourier sine series in the variable x, provided we are able to express in this way the initial conditions $\varphi(x)$, $\psi(x)$. It can be seen that such an expansion makes sense for a sufficiently wide class of functions. In such a case, we use orthogonality of the functions $\sin \frac{n\pi x}{l}$, $n = 1, 2, \ldots$, to calculate the coefficients A_n, B_n. We obtain the following expressions:

$$A_n = \frac{2}{l} \int_0^l \varphi(x) \sin \frac{n\pi x}{l}\, dx,$$

$$B_n = \frac{l}{n\pi c} \frac{2}{l} \int_0^l \psi(x) \sin \frac{n\pi x}{l}\, dx = \frac{2}{n\pi c} \int_0^l \psi(x) \sin \frac{n\pi x}{l}\, dx.$$

To ensure that the found solution of problem (7.13) is mathematically correct, it is necessary to prove that the series (7.23) converges. It depends on the properties of the functions φ, ψ and on the kind of the required convergence. These problems lie beyond the scope of this text and we will not deal with them.

Remark 7.4 Coefficients at the time variable in the arguments of goniometric functions in expression (7.23) (that is, the values $n\pi c/l$) are called *frequencies*. If we go back to the string which is described by problem (7.13), the corresponding frequencies take the form

$$\frac{n\pi \sqrt{T}}{l\sqrt{\rho}} \quad \text{for } n = 1, 2, 3, \ldots$$

The "fundamental" tone of the string corresponds to the least of these values: $\pi\sqrt{T}/(l\sqrt{\rho})$. Higher (aliquot) tones are then exactly the integer multiples of this basic tone. The discovery that musical tones can be described in this easy mathematical way was made by Euler in 1749.

Example 7.5 Let us solve the initial boundary value problem

(7.24)
$$\begin{cases} u_{tt} = c^2 u_{xx}, & x \in (0, \pi),\ t > 0, \\ u(0, t) = u(\pi, t) = 0, \\ u(x, 0) = \sin 2x, & u_t(x, 0) = 0. \end{cases}$$

Using the above relations, we can write the solution as

$$u(x, t) = \sum_{n=1}^{\infty} (A_n \cos nct + B_n \sin nct) \sin nx.$$

The zero initial velocity implies zero "sine coefficients" $B_n = 0$ for all $n \in \mathbb{N}$. The initial displacement determines the "cosine coefficients" A_n:

$$u(x,0) = \sin 2x = \sum_{n=1}^{\infty} A_n \sin nx.$$

Since the functions $\sin nx$ form a complete orthogonal set on $(0, \pi)$, we easily obtain

$$A_2 = 1, \quad A_n = 0 \quad \text{for } n \in \mathbb{N} \setminus \{2\}.$$

Hence, we can conclude that the solution of (7.24) reduces to

$$u(x,t) = \cos 2ct \sin 2x.$$

□

Example 7.6 Similarly to the previous example, we can show that the function

(7.25)
$$u(x,t) = \sum_{n=1}^{\infty} a_n \sin \frac{n\pi x}{l} \cos \frac{n\pi ct}{l},$$

with

$$a_n = \frac{2}{l} \int_0^l e^{-(x-l/2)^2} \sin \frac{n\pi x}{l}\, dx,$$

solves the initial boundary value problem

(7.26)
$$\begin{cases} u_{tt} = c^2 u_{xx}, & x \in (0, l),\ t > 0, \\ u(0, t) = u(l, t) = 0, \\ u(x, 0) = e^{-(x-l/2)^2}, & u_t(x, 0) = 0. \end{cases}$$

Figure 7.7 sketches a partial sum of the series (7.25) up to the term $n = 15$ with values $c = 6$ and $l = 8$.

□

Dirichlet Boundary Conditions, Diffusion Equation. Now let us consider an initial boundary value problem for the heat equation

(7.27)
$$\boxed{\begin{array}{l} u_t = k u_{xx}, \quad 0 < x < l,\ t > 0, \\ u(0, t) = u(l, t) = 0, \\ u(x, 0) = \varphi(x), \end{array}}$$

which can model a thin bar whose ends are kept at a constant temperature $u = 0$. The distribution of the temperature in the bar at time $t = 0$ is given by a function $\varphi = \varphi(x)$. To find the distribution of the temperature at time $t > 0$ means to find

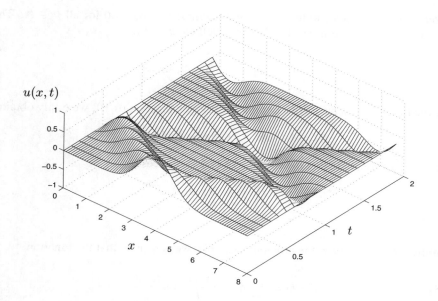

Figure 7.7 *Solution of problem (7.26).*

a solution $u = u(x, t)$ of problem (7.27). The same problem describes the diffusion process of a substance in a tube which is constructed in such a way that any substance that reaches the ends of the tube flows immediately out.

Let us proceed in the same way as in the previous example. First, we look for such a solution of the equation that satisfies only the homogeneous boundary conditions and has the form

$$u(x, t) = X(x)T(t).$$

After substitution into the heat equation and a simple arrangement, we obtain

$$-\frac{T'(t)}{kT(t)} = -\frac{X''(x)}{X(x)} = \lambda,$$

where λ is a constant. Thus, we have transformed the original equation into a couple of ODEs

(7.28) $$T' + \lambda k T = 0,$$

(7.29) $$X'' + \lambda X = 0.$$

Now, we add the homogeneous boundary conditions $X(0) = X(l) = 0$ to equation (7.29) and look for such values of λ for which this problem has a nontrivial solution. Just as in the case of the wave equation, we obtain

(7.30) $$\lambda_n = \left(\frac{n\pi}{l}\right)^2, \quad n \in \mathbb{N},$$

Section 7.2 Initial Boundary Value Problem on Finite Interval, Fourier Method

and the corresponding solutions have the form

(7.31) $$X_n(x) = C_n \sin \frac{n\pi x}{l}, \qquad n \in \mathbb{N},$$

where C_n are arbitrary constants. If we go back to equation (7.28) and substitute $\lambda = \lambda_n$, we obtain

(7.32) $$T_n(t) = A_n e^{-(n\pi/l)^2 kt}, \qquad n \in \mathbb{N}.$$

The solution u of the original problem (7.27) can be then expressed in the form of an infinite Fourier series

(7.33) $$\boxed{u(x,t) = \sum_{n=1}^{\infty} A_n e^{-(n\pi/l)^2 kt} \sin \frac{n\pi x}{l}}$$

under the assumption that the initial condition is also expandable into the corresponding series, that is,

(7.34) $$\varphi(x) = \sum_{n=1}^{\infty} A_n \sin \frac{n\pi x}{l}.$$

From the physical point of view, expression (7.33) says that, with growing time, heat is dissipated by the ends of the bar and the temperature in the whole bar decreases to zero.

Example 7.7 (Logan [14]) Let a_n be the sine Fourier coefficients of the function $\varphi(x) = 10x^3(l-x)$ on the interval $(0, l)$. Then the function

(7.35) $$u(x,t) = \sum_{n=1}^{\infty} a_n \sin \frac{n\pi x}{l} e^{-\frac{n^2 \pi^2 kt}{l^2}}$$

solves the initial boundary value problem

(7.36) $$\begin{cases} u_t = k u_{xx}, & x \in (0, l),\ t > 0, \\ u(0, t) = u(l, t) = 0, \\ u(x, 0) = 10x^3(l-x). \end{cases}$$

Figure 7.8 sketches a partial sum of the series (7.35) up to the term $n = 42$ with values $l = 1$ and $k = 1$.

□

Neumann Boundary Conditions. The same method can be used also in the case of homogeneous Neumann boundary conditions

$$u_x(0, t) = u_x(l, t) = 0.$$

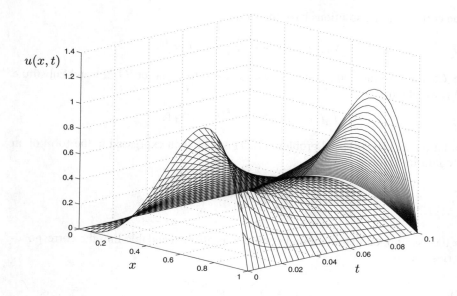

Figure 7.8 *Solution of problem (7.36).*

In the case of the wave equation, these conditions correspond to a string with free ends. If we model the diffusion process, then they describe a tube with isolated ends (nothing can penetrate in or out and the flow across the boundary is zero). Similarly, when modeling the heat flow, the homogeneous Neumann conditions represent a totally isolated tube (again, the heat flux across the boundary is zero).

Let us consider a problem for the wave or diffusion equation in the interval $(0, l)$. Separation of variables leads this time to the ODE

$$X'' + \lambda X = 0, \quad X'(0) = X'(l) = 0,$$

which has a nontrivial solution for $\lambda > 0$ and for $\lambda = 0$. In particular, we obtain

$$\lambda_n = \left(\frac{n\pi}{l}\right)^2, \quad n = 0, 1, 2, \ldots,$$

$$X_n(x) = C_n \cos \frac{n\pi x}{l}, \quad n = 0, 1, 2, \ldots.$$

The initial boundary value problem for the diffusion equation with Neumann boundary conditions has then a solution in the form

(7.37) $$\boxed{u(x,t) = \frac{1}{2}A_0 + \sum_{n=1}^{\infty} A_n e^{-(n\pi/l)^2 kt} \cos \frac{n\pi x}{l},}$$

Section 7.2 Initial Boundary Value Problem on Finite Interval, Fourier Method

provided the initial condition is expandable into the Fourier cosine series

$$\varphi(x) = \frac{1}{2}A_0 + \sum_{n=1}^{\infty} A_n \cos \frac{n\pi x}{l}.$$

In the case of an initial boundary value problem for the wave equation with homogeneous Neumann boundary conditions on $(0, l)$ we obtain

(7.38) $$u(x,t) = \frac{1}{2}A_0 + \frac{1}{2}B_0 t + \sum_{n=1}^{\infty} \left(A_n \cos \frac{n\pi ct}{l} + B_n \sin \frac{n\pi ct}{l} \right) \cos \frac{n\pi x}{l},$$

where the initial data have to satisfy

$$\varphi(x) = \frac{1}{2}A_0 + \sum_{n=1}^{\infty} A_n \cos \frac{n\pi x}{l},$$

$$\psi(x) = \frac{1}{2}B_0 + \sum_{n=1}^{\infty} \frac{n\pi c}{l} B_n \cos \frac{n\pi x}{l}.$$

In both cases, the reader is asked to carry out detailed calculations.

Example 7.8 Let a_n be the cosine Fourier coefficients of the function $\varphi(x) = \frac{1}{2} - 200x^4(1-x)^4$ in the interval $(0, l)$. Then the function

(7.39) $$u(x,t) = \frac{a_0}{2} + \sum_{n=1}^{\infty} a_n \cos \frac{n\pi x}{l} e^{-\frac{n^2\pi^2 kt}{l^2}}$$

solves the initial boundary value problem

(7.40) $$\begin{cases} u_t = k u_{xx}, & x \in (0, l), \ t > 0, \\ u_x(0, t) = u_x(l, t) = 0, \\ u(x, 0) = \frac{1}{2} - 200x^4(1-x)^4. \end{cases}$$

Figure 7.9 sketches a partial sum of the series (7.39) up to the term $n = 15$ with values $l = 1$ and $k = 1$. □

Robin Boundary Conditions. Together with the wave equation, the Robin boundary conditions describe a string whose ends are catched by springs (obeying Hooke's law) which pull it back to the equilibrium. In the case of the heat flow in a bar, these boundary conditions model the heat transfer between the bar ends and the surrounding media.

Let us consider the following modeling problem illustrated in Figure 7.10. Let us take a vertical bar of unit length, whose upper end is kept at zero temperature

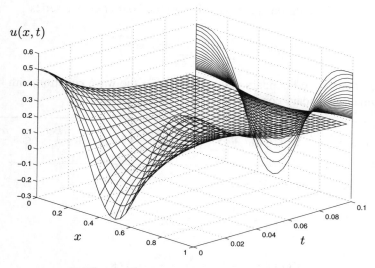

Figure 7.9 *Solution of problem (7.40).*

while the lower end is immersed into a reservoir with water of zero temperature. The heat convection proceeding between the lower end and water is described by the law $u_x(1,t) = -hu(1,t)$. The constant h corresponds to the heat transfer coefficient. Let us suppose that the initial temperature of the bar is given by a function $u(x,0) = x$. To look for the distribution of the temperature in the bar means to solve the initial boundary value problem for the heat equation with mixed boundary conditions (Dirichlet and Robin boundary conditions):

(7.41)
$$\begin{cases} u_t = k u_{xx}, & 0 < x < 1, \ t > 0, \\ u(0,t) = 0, \\ u_x(1,t) + hu(1,t) = 0, \\ u(x,0) = x. \end{cases}$$

Searching for the solution, we proceed in the same way as in the previous examples. In the first step, we separate the variables, that means, we consider a solution in the form $u(x,t) = X(x)T(t)$, and after substitution into the equation, we obtain

$$-\frac{T'}{kT} = -\frac{X''}{X} = \lambda,$$

where λ is a so far unknown constant. Thus we have transformed the original equation into a couple of independent ODEs

(7.42) $$T' + \lambda k T = 0,$$
(7.43) $$X'' + \lambda X = 0.$$

Section 7.2 Initial Boundary Value Problem on Finite Interval, Fourier Method

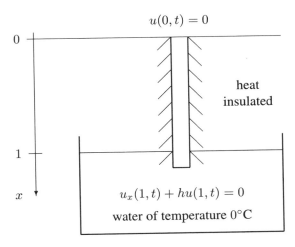

Figure 7.10 *Schematic illustration of problem (7.41).*

Moreover, the function $X(x)$ must satisfy the boundary conditions

$$X(0) = 0, \quad X'(1) + hX(1) = 0.$$

By a simple discussion we exclude the values $\lambda \leq 0$ that lead only to the trivial solution $X(x) \equiv 0$. Thus, let us consider $\lambda := \mu^2 > 0$. Then

$$X(x) = A \sin \mu x + B \cos \mu x,$$

and after substituting into the boundary conditions we obtain

$$B = 0, \quad A\mu \cos \mu + hA \sin \mu = 0.$$

Since we look for the nontrivial solution, i.e. $A \neq 0$, the last equality can be written in the form

(7.44) $$\tan \mu = -\frac{\mu}{h}.$$

To find the roots of the transcendent equation (7.44) means to find the intersections of the graphs of functions $\tan \mu$ and $-\frac{\mu}{h}$ (see Figure 7.11).

It is evident that there exists an infinite sequence of positive values μ_n, $n \in \mathbb{N}$, such that the corresponding solutions assume the form

$$X_n(x) = A_n \sin \mu_n x.$$

The following table specifies the first five approximate values μ_n for $h = 1$.

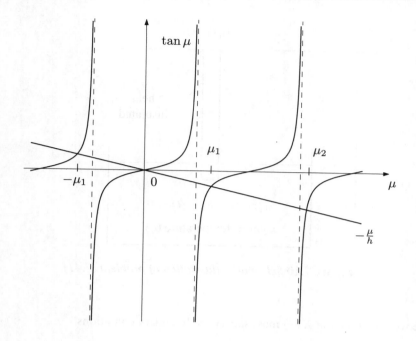

Figure 7.11 *Intersections of graphs of functions* $\tan \mu$ *and* $-\frac{\mu}{h}$.

n	1	2	3	4	5
μ_n	2.02	4.91	7.98	11.08	14.20

If we go back to equation (7.42), we obtain

$$T_n(t) = C_n e^{-k\mu_n^2 t}$$

and hence

$$u_n(x,t) = A_n e^{-k\mu_n^2 t} \sin \mu_n x.$$

The final result of the original problem is then searched in the form of the Fourier series

(7.45) $$u(x,t) = \sum_{n=1}^{\infty} A_n e^{-k\mu_n^2 t} \sin \mu_n x.$$

The coefficients A_n shall be determined in such a way that the initial condition holds, that is,

$$u(x,0) = \sum_{n=1}^{\infty} A_n \sin \mu_n x = x.$$

Section 7.2 Initial Boundary Value Problem on Finite Interval, Fourier Method

If we multiply this relation by a function $\sin \mu_m x$ and integrate from 0 to 1, we arrive at

$$\int_0^1 x \sin \mu_m x \, dx = \sum_{n=1}^{\infty} A_n \int_0^1 \sin \mu_n x \sin \mu_m x \, dx.$$

Since

$$\int_0^1 \sin \mu_n x \sin \mu_m x \, dx = \begin{cases} 0, & n \neq m, \\ \dfrac{\mu_n - \sin \mu_n \cos \mu_n}{2\mu_n}, & n = m, \end{cases}$$

we obtain, after simplification,

$$A_n = \frac{2\mu_n}{\mu_n - \sin \mu_n \cos \mu_n} \int_0^1 x \sin \mu_n x \, dx.$$

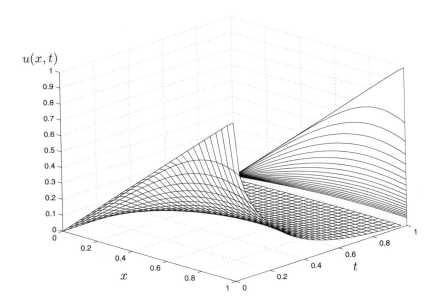

Figure 7.12 *Graphic illustration of the solution $u(x,t)$ of problem (7.41) for $h = 1$, $k = 1$.*

Remark 7.9 In the previous examples we have met special types of the boundary value problem for the second order ODE:

(7.46)
$$\begin{array}{l} -y'' = \lambda y, \quad 0 < x < l, \\ \alpha_0 y(0) + \beta_0 y'(0) = 0, \\ \alpha_1 y(l) + \beta_1 y'(l) = 0. \end{array}$$

We say that any value of the parameter $\lambda \in \mathbb{R}$ for which problem (7.46) has a nontrivial solution, is called an *eigenvalue*. The corresponding nontrivial solution is called an *eigenfunction* related to the eigenvalue λ. In the Fourier method we used some special properties of eigenvalues λ_n and eigenfunctions $y_n(x)$ of (7.46), namely that $y_n(x)$ form a complete orthogonal set. It means

$$\int_0^l y_n(x) y_m(x)\, \mathrm{d}x = 0$$

provided $y_m(x)$ and $y_n(x)$ are two eigenfunctions corresponding to two different eigenvalues λ_m and λ_n. Moreover, many functions defined on $(0, l)$ are expandable into Fourier series with respect to the eigenfunctions y_n:

$$f(x) = \sum_{n=1}^{\infty} F_n y_n(x),$$

where F_n are the Fourier coefficients defined by the relation

$$F_n = \frac{\int_0^l f(x) y_n(x)\, \mathrm{d}x}{\int_0^l y_n^2(x)\, \mathrm{d}x}.$$

It can be seen that these properties are typical not only for sines and cosines, which solve problem (7.46), but also for more general functions which arise as solutions of the so called *Sturm-Liouville boundary value problem*. The reader can find basic facts of Sturm-Liouville theory in Appendix 14.1.

The Principle of the Fourier Method for initial boundary value problems on a finite interval with a homogeneous equation and homogeneous boundary conditions can be summarized into the following steps:

(i) We search for the solution in the separated form $u(x, t) = X(x)T(t)$.

(ii) We transform the PDE into a couple of ODEs for unknown functions $X(x)$ and $T(t)$.

(iii) Considering the ODE for $X(x)$ with homogeneous boundary conditions we find the eigenvalues λ_n and the eigenfunctions $X_n(x)$ of the corresponding boundary value problem.

(iv) We substitute the eigenvalues λ_n into the ODE for the unknown function $T(t)$ and find its general solution.

(v) We write the solution of the original PDE in the form of an infinite Fourier series
$$u(x,t) = \sum_{n=1}^{\infty} X_n(x) T_n(t).$$

(vi) We expand the initial conditions into Fourier series with respect to the system of orthogonal eigenfunctions $X_n(x)$.

(vii) Comparing the expansions of the initial conditions and the solution $u(x,t)$, we calculate the remaining coefficients.

The above mentioned technique is formal and the precise justification of particular steps requires much of mathematical calculus that lies beyond the scope of this text. The interested reader can find the details, e.g., in [22].

7.3 Fourier Method for Nonhomogeneous Problems

Nonhomogeneous Equation. Let us consider the general initial boundary value problem for a nonhomogeneous heat equation

(7.47)
$$\boxed{\begin{aligned} u_t - k u_{xx} &= f(x,t), \quad 0 < x < l, \ t > 0, \\ \alpha_0 u(0,t) + \beta_0 u_x(0,t) &= 0, \\ \alpha_1 u(l,t) + \beta_1 u_x(l,t) &= 0, \\ u(x,0) &= \varphi(x). \end{aligned}}$$

The solving procedure of this problem is—in a certain sense—similar to the method of variation of parameters, which is used for solving ODEs. In the case of a homogeneous heat equation ($f(x,t) \equiv 0$), the solution of problem (7.47) is

$$u(x,t) = \sum_{n=1}^{\infty} A_n e^{-\lambda_n k t} X_n(x),$$

where $X_n(x)$ are eigenfunctions with eigenvalue λ_n of the Sturm-Liouville problem

$$\begin{cases} X'' + \lambda X = 0, \\ \alpha_0 X(0) + \beta_0 X'(0) = 0, \\ \alpha_1 X(l) + \beta_1 X'(l) = 0, \end{cases}$$

that is, problem (7.46) with $p(x) \equiv 1$, $q(x) \equiv 0$, $r(x) \equiv 1$. Now, it is natural to ask whether the solution of the nonhomogeneous problem could be expressed in the form of a more general series

$$u(x,t) = \sum_{n=1}^{\infty} T_n(t) X_n(x).$$

This reasoning has also its physical justification. Since the function $f(x,t)$ describes the density of distribution of heat sources inside the bar, we can expect that the development of temperature in time will not be expressed by terms $e^{-\lambda_n k t}$ (as in the case of the homogeneous equation, when there is no source inside the bar), but by means of some other functions $T_n(t)$ depending on $f(x,t)$.

We illustrate the process on a simple example.

Example 7.10 Let us solve the initial boundary value problem

(7.48)
$$\begin{cases} u_t - k u_{xx} = f(x,t), & 0 < x < 1,\ t > 0, \\ u(0,t) = u(1,t) = 0, \\ u(x,0) = \varphi(x). \end{cases}$$

First of all, we determine the eigenvalues λ_n and the system of eigenfunctions $X_n(x)$, $n \in \mathbb{N}$. We obtain them in the same way as in the previous section, by solving the corresponding homogeneous problem

$$\begin{cases} u_t - k u_{xx} = 0, & 0 < x < 1,\ t > 0, \\ u(0,t) = u(1,t) = 0, \\ u(x,0) = \varphi(x). \end{cases}$$

Let us recall that, in this case, we have

$$\lambda_n = (n\pi)^2, \quad X_n(x) = \sin n\pi x, \quad n \in \mathbb{N}.$$

Now, we expand all data of the original problem (7.48) into a Fourier series with respect to the eigenfunctions $X_n(x)$. That is, for a fixed $t > 0$, we write the right-hand side $f(x,t)$ as

$$f(x,t) = \sum_{n=1}^{\infty} f_n(t) \sin n\pi x,$$

where the components $f_n(t)$ are the Fourier sine coefficients of $f(x,t)$:

$$f_n(t) = 2 \int_0^1 f(x,t) \sin n\pi x \, dx.$$

Similarly, we expand the initial condition to

$$\varphi(x) = \sum_{n=1}^{\infty} \varphi_n \sin n\pi x$$

with

$$\varphi_n = 2 \int_0^1 \varphi(x) \sin n\pi x \, dx.$$

Section 7.3 Fourier Method for Nonhomogeneous Problems

Now, we search for the solution of problem (7.48) in the form of a series

$$u(x,t) = \sum_{n=1}^{\infty} T_n(t) \sin n\pi x.$$

Substituting all the above expansions into the equation of (7.48), we obtain

(7.49) $$\sum_{n=1}^{\infty} T'_n(t) \sin n\pi x + k \sum_{n=1}^{\infty} (n\pi)^2 T_n(t) \sin n\pi x = \sum_{n=1}^{\infty} f_n(t) \sin n\pi x.$$

Now, we multiply this equality by $\sin n\pi x$ and integrate over $(0,1)$ in x. Due to the completeness of the system of functions $\sin n\pi x$, (7.49) is equivalent to the system of ODEs

$$T'_n + k(n\pi)^2 T_n = f_n(t), \quad n \in \mathbb{N}.$$

To fulfil the initial condition

$$u(x,0) = \sum_{n=1}^{\infty} T_n(0) \sin n\pi x = \sum_{n=1}^{\infty} \varphi_n \sin n\pi x = \varphi(x),$$

all functions $T_n(t)$ must satisfy

$$T_n(0) = \varphi_n = 2 \int_0^1 \varphi(x) \sin n\pi x \, dx.$$

Hence, we easily obtain

$$T_n(t) = \varphi_n e^{-k(n\pi)^2 t} + \int_0^t e^{-k(n\pi)^2(t-\tau)} f_n(\tau) \, d\tau.$$

The resulting solution of problem (7.48) then assumes the form

$$u(x,t) = \sum_{n=1}^{\infty} \varphi_n e^{-k(n\pi)^2 t} \sin n\pi x + \sum_{n=1}^{\infty} \sin n\pi x \int_0^t e^{-k(n\pi)^2(t-\tau)} f_n(\tau) \, d\tau.$$

The first sum on the right-hand side represents the solution corresponding to the homogeneous problem with the given initial condition, whereas the other one describes the influence of the right-hand side. □

Nonhomogeneous Boundary Conditions and Their Transformation. As we could see in the previous paragraphs, the assumption of homogeneity of boundary conditions is essential for applicability of the Fourier method. That is why we will now study the problem of transforming the initial boundary value problem with nonhomogeneous boundary conditions to a problem with homogeneous conditions.

Let us consider the following model situation. Let the heat-insulated bar of length l have its ends kept at constant temperatures g_0 and g_1. The initial temperature distribution is given by a function $\varphi = \varphi(x)$. The development of the temperature $u(x,t)$ in the bar is thus the solution of the initial boundary value problem

(7.50)
$$\begin{cases} u_t - ku_{xx} = 0, & 0 < x < l, \ t > 0, \\ u(0,t) = g_0, \quad u(l,t) = g_1, \\ u(x,0) = \varphi(x). \end{cases}$$

Physical intuition leads us to the hypothesis that, for $t \to \infty$, the distribution of temperature $u(x,t)$ in the bar converges to the linear function $w(x) = g_0 + (g_1 - g_0)x/l$. It is thus reasonable to assume that the solution of problem (7.50) will have the form

(7.51)
$$u(x,t) = \left(g_0 + \frac{x}{l}(g_1 - g_0)\right) + U(x,t).$$

Here the term $w(x) = g_0 + (g_1 - g_0)x/l$ represents the stationary part (it does not depend on time and satisfies the equation $w_t = kw_{xx}$ and the boundary conditions $w(0) = g_0$, $w(1) = g_1$). The term $U(x,t)$ represents the time-dependent part, which converges to zero for $t \to \infty$. Due to the fact that the stationary part $w(x)$ is uniquely determined by the constants g_0, g_1, we can—instead of the function $u(x,t)$—look directly for the unknown function $U(x,t)$. If we insert expression (7.51) into (7.50), we find out that $U(x,t)$ solves the *homogeneous* initial boundary value problem

$$\begin{cases} U_t - kU_{xx} = 0, & 0 < x < l, \ t > 0, \\ U(0,t) = 0, \quad U(l,t) = 0, \\ U(x,0) = \varphi(x) - [g_0 + \frac{x}{l}(g_1 - g_0)] =: \bar{\varphi}(x), \end{cases}$$

which can be solved by the standard Fourier method.

In practice, however, we have more often to deal with boundary conditions that *depend on time*. We illustrate their transformation to the homogeneous boundary conditions on an example. Let us consider the initial boundary value problem

(7.52)
$$\begin{cases} u_t - ku_{xx} = 0, & 0 < x < l, \ t > 0, \\ u(0,t) = g_1(t), \\ u_x(l,t) + hu(l,t) = g_2(t), \\ u(x,0) = \varphi(x). \end{cases}$$

The solution will be searched for in the form

$$u(x,t) = A(t)\left(1 - \frac{x}{l}\right) + B(t)\frac{x}{l} + U(x,t),$$

where functions $A(t)$ and $B(t)$ will be chosen so that the "quasi-stationary" part $w(x,t) = A(t)\left(1 - \frac{x}{l}\right) + B(t)\frac{x}{l}$ will satisfy the boundary conditions of problem (7.52).

The function $U(x,t)$ must then fulfil the homogeneous boundary conditions. If we substitute the function $w(x,t)$ into the boundary conditions

$$w(0,t) \equiv A(t) = g_1(t),$$
$$w_x(l,t) + hw(l,t) \equiv -\frac{A(t)}{l} + \frac{B(t)}{l} + hB(t) = g_2(t),$$

we obtain

$$A(t) = g_1(t), \quad B(t) = \frac{g_1(t) + lg_2(t)}{1 + lh}$$

and thus

$$u(x,t) = g_1(t)\left(1 - \frac{x}{l}\right) + \frac{g_1(t) + lg_2(t)}{1 + lh}\frac{x}{l} + U(x,t).$$

Substituting this expression into (7.52), we easily find out that $U(x,t)$ must solve the initial boundary value problem

$$\begin{cases} U_t - kU_{xx} = -w_t, & 0 < x < l,\ t > 0, \\ U(0,t) = 0, \\ U_x(l,t) + hU(l,t) = 0, \\ U(x,0) = \varphi(x) - w(x,0). \end{cases}$$

In this case we have transformed the original problem with a homogeneous equation and nonhomogeneous boundary conditions into a problem with nonzero right-hand side but with homogeneous boundary conditions, which can be solved by the Fourier method.

7.4 Transformation to Simpler Problems

The goal of this section is to point out some transformations that can lead to simpler PDEs.

Example 7.11 (Lateral Heat Transfer in the Bar.) Let us consider the initial boundary value problem

(7.53)
$$\begin{array}{|l|} \hline u_t - ku_{xx} + qu = 0, \quad 0 < x < 1,\ t > 0, \\ u(0,t) = u(1,t) = 0, \\ u(x,0) = \varphi(x). \\ \hline \end{array}$$

This problem describes heat conduction in the bar, where the heat is transferred to the surroundings by the bar surface. The heat-transfer coefficient is denoted by q. We look for a substitution that would simplify the PDE of problem (7.53). We will base our considerations on the physical properties of the model. The temperature $u(x,t)$ develops at every point x_0 in terms of the following two phenomena:

1. the diffusion of the heat along the bar (described by the term $-ku_{xx}$),

2. the heat transfer by the lateral bar surface (described by the term qu).

Let us assume that there is no diffusion along the bar (that is, $k = 0$). Then the development of temperature at every point of the bar is given by

$$u(x_0, t) = u(x_0, 0) e^{-qt}$$

(since for $k = 0$ the function $u(x_0, t)$ solves the ODE $u_t + qu = 0$ with the initial condition $u = u(x_0, 0)$). Making use of this fact, we try to express the solution of the initial boundary value problem (7.53) (now, with $k \neq 0$) in the form

(7.54) $$u(x, t) = e^{-qt} w(x, t),$$

where $w = w(x, t)$ describes the heat transfer caused only by the diffusion process. If we substitute (7.54) into (7.53), we obtain the following problem for the required function $w(x, t)$:

$$\begin{cases} w_t - k w_{xx} = 0, & 0 < x < 1,\ t > 0, \\ w(0, t) = w(1, t) = 0, \\ w(x, 0) = \varphi(x). \end{cases}$$

This is nothing but the classical homogeneous problem for the heat equation, the solution of which is already known to the reader.

□

Example 7.12 (Problem with Convective Term.) The PDE

(7.55) $$\boxed{u_t - k u_{xx} + v u_x = 0}$$

describes the so called *convective diffusion*, where v is a constant representing the propagation speed of the medium (see Section 5.1). Equation (7.55) can be transformed into the standard diffusion equation by the substitution

$$u(x, t) = e^{\frac{v}{2k}(x - \frac{vt}{2})} w(x, t).$$

(The reader is asked to verify it.) The exponential term in this case reflects the motion of the medium, $w(x, t)$ corresponds only to the diffusion process.

□

7.5 Exercises

1. Prove that if $f(x) \in C^2([0, \infty))$, then its odd extension $\tilde{f}(x)$ is of the class $C^2(\mathbb{R})$ if and only if $f(0) = f''(0) = 0$.

2. Using the method of even extension, derive the formula for the solution of the diffusion equation on the half-line with homogeneous Neumann boundary condition at $x = 0$. Consider the general initial condition $u(x, 0) = \varphi(x)$.

$$[u(x, t) = \int_0^\infty \varphi(y)(G(x + y, t) + G(x - y, t)) dy]$$

Section 7.5 Exercises

3. Using the method of even extension, derive the formula for the solution of the wave equation on the half-line with homogeneous Neumann boundary condition at $x = 0$. Consider the general initial conditions $u(x,0) = \varphi(x)$, $u_t(x,0) = \psi(x)$.

$$[u(x,t) = \tfrac{1}{2}(\varphi(x+ct) + \varphi(x-ct)) + \tfrac{1}{2c}\int_{x-ct}^{x+ct} \psi(\tau)d\tau \quad \text{for } x > ct,$$

$$u(x,t) = \tfrac{1}{2}(\varphi(ct+x) + \varphi(ct-x)) + \tfrac{1}{2c}\int_0^{ct+x} \psi(\tau)d\tau + \tfrac{1}{2c}\int_0^{ct-x} \psi(\tau)d\tau \quad \text{for } 0 < x < ct]$$

4. Find a solution of the problem

$$\begin{cases} u_{tt} = u_{xx}, & x > 0,\ t > 0, \\ u(0,t) = 0, \\ u(x,0) = 1,\ u_t(x,0) = 0. \end{cases}$$

Sketch the graph of the solution on several time levels.

$[u(x,t) = 1 \text{ for } x > t,\ u(x,t) = 0 \text{ for } 0 < x < t]$

5. Find a solution of the problem

$$\begin{cases} u_{tt} = u_{xx}, & x > 0,\ t > 0, \\ u(0,t) = 0, \\ u(x,0) = xe^{-x},\ u_t(x,0) = 0. \end{cases}$$

Sketch the graph of the solution on several time levels. Notice the wave reflection on the boundary.

$$[u(x,t) = \tfrac{1}{2}(x+t)e^{-x-t} + \tfrac{1}{2}(x-t)e^{-x+t} \quad \text{for } x > t,$$

$$u(x,t) = \tfrac{1}{2}(t+x)e^{-t-x} - \tfrac{1}{2}(t-x)e^{-t+x} \quad \text{for } 0 < x < t]$$

6. Find a solution of the problem

$$\begin{cases} u_{tt} = u_{xx}, & x > 0,\ t > 0, \\ u(0,t) = 0, \\ u(x,0) = \varphi(x),\ u_t(x,0) = 0, \end{cases}$$

with

$$\varphi(x) = \begin{cases} \cos^3 x, & x \in (\tfrac{3\pi}{2}, \tfrac{5\pi}{2}), \\ 0, & x \in (0, \infty) \setminus (\tfrac{3\pi}{2}, \tfrac{5\pi}{2}). \end{cases}$$

Sketch the graph of the solution on several time levels.

7. Find a solution of the problem

$$\begin{cases} u_t = ku_{xx}, & x > 0,\ t > 0, \\ u(0,t) = 1, \\ u(x,0) = 0. \end{cases}$$

$[u(x,t) = 0 \text{ for } x > t,\ u(x,t) = 1 \text{ for } 0 < x < t]$

8. The heat flow in a metal rod with an inner heat source is described by the problem
$$\begin{cases} u_t = ku_{xx} + 1, & 0 < x < l, \ t > 0, \\ u(0,t) = 0, & u(1,t) = 1. \end{cases}$$

What will be the temperature of the rod in the steady state that will be achieved after a sufficiently long time? (Realize that in the steady state u depends only on x.) Does the absence of an initial condition cause any trouble?

$[u(x) = -\frac{x^2}{2k} + (1+\frac{1}{2k})x]$

9. Consider the case that the heat leaks from the rod over its lateral surface at a speed proportional to its temperature u. The corresponding problem has the form
$$\begin{cases} u_t = ku_{xx} - au, & 0 < x < l, \ t > 0, \\ u(0,t) = 0, & u(l,t) = 1. \end{cases}$$

Draw the temperature distribution in the steady state and discuss how the heat flows in the rod and across its boundary.

10. Bacteria in a one-dimensional medium (a tube of a unit cross-section, length l, closed on both ends) breed according to the logistic law $ru(l - u/K)$, where r is a growth constant, K is the capacity of the medium, and $u = u(x,t)$ denotes the density of bacteria per unit length. At the beginning, the density is given by $u = ax(l-x)$. At time $t > 0$, the bacteria also diffuse with the diffusion constant D. Formulate the initial boundary value problem describing their density. What will be the density distribution if we wait long enough? Sketch intuitively several profiles illustrating the density evolution in time. Consider cases $al^2 < 4K$ and $al^2 > 4K$ separately.

11. Consider a bar of length l which is insulated in such a way that there is no exchange of heat with the surrounding medium. Show that the average temperature
$$\frac{1}{l}\int_0^l u(x,t)\,dx$$
is constant with respect to time t.

[Hint: Integrate the corresponding series term by term.]

12. Solve the problem describing the motion of a string of unit length
$$\begin{cases} u_{tt} = c^2 u_{xx}, & 0 < x < 1, \ t > 0, \\ u(0,t) = u(1,t) = 0, \\ u(x,0) = \varphi(x), \ u_t(x,0) = \psi(x) \end{cases}$$

for the data given below. Illustrate the string motion by a graphic representation of a partial sum of the resulting series for various values t. Comparing the graph for $t = 0$ and the graph of the function $\varphi(x)$, decide whether the number of terms in the sum is sufficient.

(a) $\varphi(x) = 0.05\sin \pi x$, $\psi(x) = 0$, $c = 1/\pi$.

$$[u(x,t) = 0.05\sin \pi x \cos t]$$

(b) $\varphi(x) = \sin \pi x \cos \pi x$, $\psi(x) = 0$, $c = 1/\pi$.

(c) $\varphi(x) = \sin \pi x + 3\sin 2\pi x - \sin 5\pi x$, $\psi(x) = 0$, $c = 1$.

$$[u(x,t) = \sin \pi x \cos \pi t + 3\sin 2\pi x \cos 2\pi t - \sin 5\pi x \cos 5\pi t]$$

(d) $\varphi(x) = \sin \pi x + 0.5\sin 3\pi x + 3\sin 7\pi x$, $\psi(x) = \sin 2\pi x$, $c = 1$.

(e) $\psi(x) = 0$, $c = 4$,

$$\varphi(x) = \begin{cases} 2x, & 0 \le x \le 1/2, \\ 2(1-x), & 1/2 < x \le 1. \end{cases}$$

$$[u(x,t) = \sum_{k=0}^{\infty} \frac{8(-1)^k}{\pi^2(2k+1)^2} \sin(2k+1)\pi x \cos 4(2k+1)\pi t]$$

(f) $\psi(x) = 2$, $c = 1/\pi$,

$$\varphi(x) = \begin{cases} 0, & 0 \le x \le 1/3, \\ 1/30(x - 1/3), & 1/3 \le x \le 2/3, \\ 1/30(1-x), & 2/3 < x \le 1. \end{cases}$$

(g) $\psi(x) = 1$, $c = 4$,

$$\varphi(x) = \begin{cases} 4x, & 0 \le x \le 1/4, \\ 1, & 1/4 \le x \le 3/4, \\ 4(1-x), & 3/4 < x \le 1. \end{cases}$$

$$[u(x,t) = \sum_{n=1}^{\infty} \frac{8}{\pi^2 n^2} (\sin(n\pi/4) + \sin(3n\pi/4)) \sin n\pi x \cos 4n\pi t$$

$$+ \sum_{k=0}^{\infty} \frac{1}{\pi^2(2k+1)^2} \sin(2k+1)\pi x \sin 4(2k+1)\pi t]$$

(h) $\varphi(x) = x\sin \pi x$, $\psi(x) = 0$, $c = 1/\pi$.

(i) $\varphi(x) = x(1-x)$, $\psi(x) = \sin \pi x$, $c = 1$.

$$[u(x,t) = \sum_{k=1}^{\infty} \frac{8}{\pi^3(2k+1)^3} \sin(2k+1)\pi x \cos(2k+1)\pi t + \frac{1}{\pi}\sin \pi x \sin \pi t]$$

(j) $\psi(x) = 0$, $c = 1$,

$$\varphi(x) = \begin{cases} 4x, & 0 \le x \le 1/4, \\ -4(x - 1/2), & 1/4 \le x \le 3/4, \\ 4(x-1), & 3/4 < x \le 1. \end{cases}$$

13. Solve the wave equation on the interval $(0, 4\pi)$ with $c = 1$, homogeneous Dirichlet boundary conditions, zero initial velocity $\psi = 0$ and the initial displacement given by

$$\varphi(x) = \begin{cases} \cos^3 x, & x \in [\frac{3\pi}{2}, \frac{5\pi}{2}], \\ 0, & x \in [0, 4\pi] \setminus [\frac{3\pi}{2}, \frac{5\pi}{2}]. \end{cases}$$

Plot the graph of the solution on several time levels.

14. Solve the wave equation on the interval $(0, 4\pi)$ with $c = 1$, homogeneous Neumann boundary conditions, zero initial velocity $\psi = 0$ and the initial displacement given by
$$\varphi(x) = \begin{cases} \cos^3 x, & x \in [\frac{3\pi}{2}, \frac{5\pi}{2}], \\ 0, & x \in [0, 4\pi] \setminus [\frac{3\pi}{2}, \frac{5\pi}{2}]. \end{cases}$$
Plot the graph of the solution on several time levels.

15. Solve the following initial boundary value problem for the wave equation:
$$\begin{cases} u_{tt} - u_{xx} = 0, & 0 < x < 3,\ t > 0, \\ u(0, t) = 0, & u(3, t) = 0, \\ u(x, 0) = 1 - \cos \frac{\pi x}{3}, \\ u_t(x, 0) = 0. \end{cases}$$

16. Solve the following initial boundary value problem for the wave equation:
$$\begin{cases} u_{tt} - u_{xx} = 0, & 0 < x < 1,\ t > 0, \\ u(0, t) = 0, & u(1, t) = 0, \\ u(x, 0) = x \cos \pi x, \\ u_t(x, 0) = 1. \end{cases}$$

17. A string of length 2π with fixed ends is excited by the impact of a hammer which gives it the initial velocity
$$\psi(x) = \begin{cases} 100, & \frac{\pi}{2} < x < \frac{3\pi}{2}, \\ 0, & x \in [0, 2\pi] \setminus (\frac{\pi}{2}, \frac{3\pi}{2}). \end{cases}$$
Find the string vibrations provided the initial displacement was zero.

18. A uniform string with a fixed end at 0 and free end at 2π has the initial displacement
$$\varphi(x) = \begin{cases} -x, & 0 \leq x < \frac{3\pi}{2}, \\ 3x - 6\pi, & \frac{3\pi}{2} \leq x \leq 2\pi \end{cases}$$
and zero initial velocity. Assume that the string is vibrating in the medium that resists the vibrations. The resistance is proportional to the velocity with the constant of proportionality 0.03. Formulate the corresponding model and find the solution.

19. Solve the problem
$$\begin{cases} u_{tt} + u_t = u_{xx}, & 0 < x < \pi,\ t > 0, \\ u(0, t) = u(\pi, t) = 0, \\ u(x, 0) = \sin x,\ u_t(x, 0) = 0. \end{cases}$$

$[u(x, t) = e^{-t/2}(\cos \frac{\sqrt{3}}{2}t + \frac{1}{\sqrt{3}} \sin \frac{\sqrt{3}}{2}t) \sin x]$

Section 7.5 Exercises

20. Solve the problem

$$\begin{cases} u_{tt} + u_t = u_{xx}, & 0 < x < \pi, \ t > 0, \\ u(0,t) = u(\pi, t) = 0, \\ u(x,0) = x \sin x, \ u_t(x,0) = 0. \end{cases}$$

$$[u(x,t) = \tfrac{\pi}{2} e^{-t/2}(\cos \tfrac{\sqrt{3}}{2} t + \tfrac{1}{\sqrt{3}} \sin \tfrac{\sqrt{3}}{2} t) \sin x$$

$$- \tfrac{16}{\pi} e^{-t/2} \sum_{k=1}^{\infty} \tfrac{k}{(4k^2-1)^2} (\cos \sqrt{4k^2 - \tfrac{1}{4}} t + \tfrac{1}{2\sqrt{4k^2-1/4}} \sin \sqrt{4k^2 - \tfrac{1}{4}} t) \sin 2kx]$$

21. Solve the problem

$$\begin{cases} u_{tt} + 3u_t = u_{xx}, & 0 < x < \pi, \ t > 0, \\ u(0,t) = u(\pi,t) = 0, \\ u(x,0) = 0, \ u_t(x,0) = 10. \end{cases}$$

Illustrate by a graph the fact that the solution decreases to zero for t going to infinity.

$$[u(x,t) = \tfrac{16\sqrt{5}}{\pi} e^{-3t/2} \sin x \sinh \tfrac{\sqrt{5}}{2} t$$

$$+ \tfrac{40}{\pi} e^{-3t/2} \sum_{k=1}^{\infty} \tfrac{1}{(2k+1)\sqrt{(2k+1)^2 - 9/4}} \sin(2k+1)x \sin \sqrt{(2k+1)^2 - 9/4}\, t]$$

22. Solve the problem

$$\begin{cases} u_t = k u_{xx}, & 0 < x < l, \ t > 0, \\ u(0,t) = u(l,t) = 0, \\ u(x,0) = \varphi(x) \end{cases}$$

for the following data.

(a) $l = \pi$, $k = 1$, $\varphi(x) = 78$.

$$[u(x,t) = \tfrac{312}{\pi} \sum_{k=0}^{\infty} \tfrac{1}{2k+1} e^{-(2k+1)^2 t} \sin(2k+1)x]$$

(b) $l = \pi$, $k = 1$, $\varphi(x) = 30 \sin x$.

(c) $l = \pi$, $k = 1$, $\varphi(x) = \begin{cases} 33x, & 0 < x \le \pi/2, \\ 33(\pi - x), & \pi/2 < x < \pi. \end{cases}$

$$[u(x,t) = \tfrac{132}{\pi} \sum_{k=0}^{\infty} \tfrac{(-1)^k}{(2k+1)^2} e^{-(2k+1)^2 t} \sin(2k+1)x]$$

(d) $l = \pi$, $k = 1$, $\varphi(x) = \begin{cases} 100, & 0 < x \le \pi/2, \\ 0, & \pi/2 < x < \pi. \end{cases}$

(e) $l = 1$, $k = 1$, $\varphi(x) = x$.

$$[u(x,t) = \tfrac{2}{\pi} \sum_{n=1}^{\infty} \tfrac{(-1)^{n+1}}{n} e^{-n^2 \pi^2 t} \sin n\pi x]$$

(f) $l = 1$, $k = 1$, $\varphi(x) = e^{-x}$.

23. Draw the temperature distribution for various $t > 0$ for the values from Exercise 22(a). Estimate how long it takes until the maximal temperature decreases to $50°C$.

24. Solve the problem
$$\begin{cases} u_t = k u_{xx}, & 0 < x < l,\ t > 0, \\ u_x(0,t) = u_x(l,t) = 0, \\ u(x,0) = \varphi(x) \end{cases}$$
for the following data.

(a) $l = \pi$, $k = 1$, $\varphi(x) = 100$.
   [In this case, the answer can be guessed by physical intuition: $u(x,t) = 100$.]

(b) $l = \pi$, $k = 1$, $\varphi(x) = x$.

(c) $l = \pi$, $k = 1$, $\varphi(x) = \begin{cases} 100x, & 0 < x \leq \pi/2, \\ 100(\pi - x), & \pi/2 < x < \pi. \end{cases}$

$$[u(x,t) = 25\pi - \frac{200}{\pi} \sum_{n=0}^{\infty} \frac{1}{(2n+1)^2} e^{-4(2n+1)^2 t} \cos 2(2n+1)x]$$

(d) $l = 1$, $k = 1$, $\varphi(x) = \begin{cases} 100, & 0 < x \leq 1/2, \\ 0, & 1/2 < x < \pi. \end{cases}$

(e) $l = 1$, $k = 1$, $\varphi(x) = \cos \pi x$.

$$[u(x,t) = e^{-\pi^2 t} \cos \pi x]$$

(f) $l = 1$, $k = 1$, $\varphi(x) = \sin \pi x$.

25. Solve the following initial boundary value problem for the diffusion equation:
$$\begin{cases} u_t - u_{xx} = 0, & 0 < x < 2,\ t > 0, \\ u_x(0,t) = 0, & u_x(2,t) = 0, \\ u(x,0) = \varphi(x), \end{cases}$$
where
$$\varphi(x) = \begin{cases} x, & 0 \leq x \leq 1, \\ (2-x), & 1 \leq x \leq 2. \end{cases}$$

26. Solve the following initial boundary value problem for the diffusion equation:
$$\begin{cases} u_t - u_{xx} + 2u = 0, & 0 < x < 1,\ t > 0, \\ u(0,t) = 0, & u(1,t) = 0, \\ u(x,0) = \cos x. \end{cases}$$

27. Solve the nonhomogeneous initial boundary value problem
$$\begin{cases} u_t = k u_{xx}, & 0 < x < l,\ t > 0, \\ u(0,t) = T_1,\ u(l,t) = T_2, & t > 0, \\ u(x,0) = \varphi(x), & 0 < x < l \end{cases}$$
for the following data:

Section 7.5 Exercises

(a) $T_1 = 100$, $T_2 = 0$, $\varphi(x) = 30\sin(\pi x)$, $l = 1$, $k = 1$.

$$[u(x,t) = 100(1-x) + 30e^{-\pi^2 t}\sin \pi x - \tfrac{200}{\pi}\sum_{n=1}^{\infty}\tfrac{1}{n}e^{-n^2\pi^2 t}\sin n\pi x]$$

(b) $T_1 = 100$, $T_2 = 100$, $\varphi(x) = 50x(1-x)$, $l = 1$, $k = 1$.

(c) $T_1 = 100$, $T_2 = 50$, $\varphi(x) = \begin{cases} 33x, & 0 < x \leq \pi/2, \\ 33(\pi - x), & \pi/2 < x < \pi, \end{cases}$

$l = \pi$, $k = 1$.

$$[u(x,t) = 100 - \tfrac{50x}{\pi} + \tfrac{132}{\pi}\sum_{k=0}^{\infty}\tfrac{(-1)^k}{(2k+1)^2}e^{-(2k+1)^2 t}\sin(2k+1)x$$
$$-\tfrac{100}{\pi}\sum_{n=1}^{\infty}\tfrac{2-(-1)^n}{n}e^{-n^2 t}\sin nx]$$

(d) $T_1 = 0$, $T_2 = 100$, $\varphi(x) = \begin{cases} 100, & 0 < x \leq \pi/2, \\ 0, & \pi/2 < x < \pi, \end{cases}$

$l = \pi$, $k = 1$.

28. Solve the following initial boundary value problem for the diffusion equation:

$$\begin{cases} u_t - u_{xx} = 0, & 0 < x < 1,\ t > 0, \\ u(0,t) = 2, & u(1,t) = 6, \\ u(x,0) = \sin 2\pi x + 4x. \end{cases}$$

29. Solve the following initial boundary value problem for the nonhomogeneous wave equation:

$$\begin{cases} u_{tt} - 4u_{xx} = 2\sin \pi x, & 0 < x < 1,\ t > 0, \\ u(0,t) = 1, & u(1,t) = 1, \\ u(x,0) = 0, \\ u_t(x,0) = 0. \end{cases}$$

30. Solve the wave equation

$$u_{tt} - c^2 u_{xx} = A\sin \omega t, \quad 0 < x < l,\ t > 0,$$

with zero initial and boundary conditions. For which ω does the solution grow in time (the so called resonance occurs)?

31. Consider heat flow in a thin circular ring of unit radius that is insulated along its lateral surface. The temperature distribution in the ring can be described by the standard one-dimensional diffusion equation, where x represents the arc length along the ring. The shape of the domain causes that we have to consider periodic boundary conditions

$$u(-\pi, t) = u(\pi, t), \quad u_x(-\pi, t) = u_x(\pi, t).$$

Solve this problem for a general initial condition $u(x, 0) = \varphi(x)$, $x \in (-\pi, \pi)$.

32. Separate the following PDEs into appropriate ODEs:

    (a) $u_t = k u_{xx} + u$,

    (b) $u_t = k u_{xx} - m u_x + u$,

    (c) $u_t = (k(x) u_x)_x + u$.

33. Determine if the following PDEs are separable. If so, separate them into appropriate ODEs. If not, explain why.

    (a) $u_{tt} = c^2 u_{xx} + u$,

    (b) $u_{tt} = c^2 u_{xx} - m u_x + u$,

    (c) $c(x)\rho(x) u_{tt} = (T(x) u_x)_x - u_t + u$.

34. Solve the following problem:

$$\begin{cases} u_{tt} = u_{xx} - u_t + u_x, & 0 < x < \frac{3\pi}{2},\ t > 0, \\ u(0, t) = 0, \quad u(3\pi/2, t) = 0, \\ u(x, 0) = \cos x, \\ u_t(x, 0) = x^2 - (3\pi/2)^2. \end{cases}$$

Explain its physical meaning.

35. Solve the initial boundary value problem

$$\begin{cases} u_{tt} - u_{xx} + 2 u_t = \sin^3 x, & 0 < x < \pi,\ t > 0, \\ u(0, t) = 0, \quad u(\pi, t) = 0, \\ u(x, 0) = \sin x, \\ u_t(x, 0) = 0. \end{cases}$$

Here, use the identity $\sin^3 x = \frac{1}{4}(3 \sin x - \sin 3x)$.

Exercises 1, 6, 13–16, 25, 26, 28–30 are taken from Stavroulakis and Tersian [18], 8–10 from Logan [14], 11, 12, 19–24, 27, 31 from Asmar [3], and 17, 18, 32–34 from Keane [12], 35 from Barták et al. [4].

8 Solutions of Boundary Value Problems for Stationary Equations

In this chapter we consider two-dimensional boundary value problems for the Laplace (or Poisson) equation. The basic mathematical problem is to solve these equations on a given domain (open and connected set) $\Omega \subset \mathbb{R}^2$ with given conditions on the boundary $\partial\Omega$:

$$\begin{aligned}
\Delta u &= f && \text{in } \Omega, \\
u &= h_1 && \text{on } \Gamma_1, \\
\frac{\partial u}{\partial n} &= h_2 && \text{on } \Gamma_2, \\
\frac{\partial u}{\partial n} + au &= h_3 && \text{on } \Gamma_3,
\end{aligned}$$

where f and h_i, $i = 1, 2, 3$, are given functions, a is a given constant, and $\Gamma_1 \cup \Gamma_2 \cup \Gamma_3 = \partial\Omega$. In particular, some of the boundary segments can be empty.

On a rectangle (or on a strip, on a half-plane), the solution of the Laplace equation can be found using the separation of variables (the Fourier method). The general scheme is the same as in the case of evolution equations.

1. The solution of PDE is searched in a separated form.

2. We take into account the homogeneous boundary conditions and obtain the eigenvalues of the problem. It is in this step that the geometry of the rectangle is very important.

3. The solution is written in the form of a series.

4. We include the nonhomogeneous initial or boundary conditions.

There are several special domains which can be transformed to a rectangle. For example, this is the case with the disc or its suitable parts if we use the transformation into polar coordinates.

8.1 Laplace Equation on Rectangle

Let us consider the Laplace equation $u_{xx} + u_{yy} = 0$ on a rectangle $R = \{0 < x < a,\ 0 < y < b\}$ with the boundary conditions illustrated in Figure 8.1.
We thus solve the problem

(8.1)
$$\begin{aligned}
u_{xx} + u_{yy} &= 0 \quad \text{in } R, \\
u(0, y) &= u_x(a, y) = 0, \\
u_y(x, 0) + u(x, 0) &= 0, \\
u(x, b) &= g(x).
\end{aligned}$$

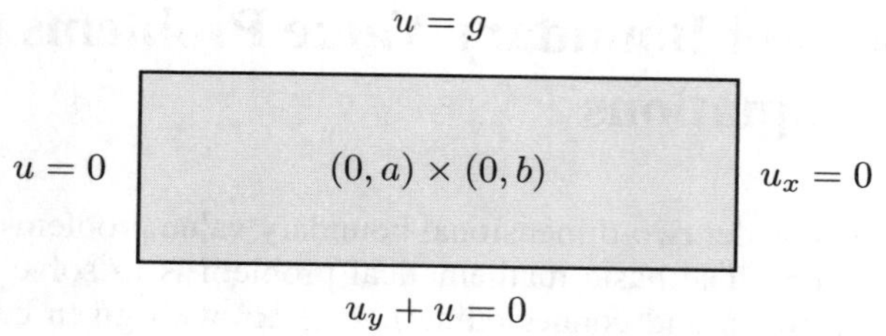

```
           u = g
     ┌─────────────────┐
u=0  │  (0,a) × (0,b)  │  u_x = 0
     └─────────────────┘
         u_y + u = 0
```

Figure 8.1 *The rectangle R and boundary conditions of (8.1).*

In the first step, we look for the solution in a separated form: $u(x,y) = X(x)Y(y)$, $X \neq 0$, $Y \neq 0$. Substituting into the equation and dividing by XY, we obtain

$$\frac{X''}{X} + \frac{Y''}{Y} = 0.$$

There must exist a constant λ such that $X'' + \lambda X = 0$ for $0 < x < a$, and $Y'' - \lambda Y = 0$ for $0 < y < b$. Moreover, the function X must fulfil the homogeneous boundary conditions $X(0) = X'(a) = 0$. By simple analysis, we find out that a nontrivial solution $X = X(x)$ exists only for

(8.2) $$\lambda_n = \beta_n^2 = \left(n + \frac{1}{2}\right)^2 \frac{\pi^2}{a^2} \qquad n = 0, 1, 2, 3, \ldots$$

and the corresponding solutions (up to a factor) are

(8.3) $$X_n(x) = \sin \frac{(n + \frac{1}{2})\pi x}{a}.$$

Now, we return to the variable y and solve the problem

$$Y'' - \beta_n^2 Y = 0, \qquad Y'(0) + Y(0) = 0$$

(the *nonhomogeneous* boundary condition for $y = b$ will be considered in the last step). Since all $\lambda_n = \beta_n^2$ are positive, we obtain Y in the form

$$Y(y) = A \cosh \beta_n y + B \sinh \beta_n y.$$

Further, we have $0 = Y'(0) + Y(0) = B\beta_n + A$. Without loss of generality, we can put $B = -1$, and thus $A = \beta_n$. Hence, we obtain

$$Y_n(y) = \beta_n \cosh \beta_n y - \sinh \beta_n y.$$

The series

(8.4) $$u(x,y) = \sum_{n=0}^{\infty} A_n \sin \beta_n x (\beta_n \cosh \beta_n y - \sinh \beta_n y)$$

then represents the harmonic function on the rectangle R that satisfies the homogeneous boundary conditions $u(0, y) = 0$, $u_x(a, y) = 0$ for $y \in (0, b)$, and $u_y(x, 0) + u(x, 0) = 0$ for $x \in (0, a)$. It remains to deal with the boundary condition $u(x, b) = g(x)$. In order to satisfy it, we must ensure that

$$(8.5) \qquad g(x) = \sum_{n=0}^{\infty} A_n (\beta_n \cosh \beta_n b - \sinh \beta_n b) \sin \beta_n x$$

for all $x \in (0, a)$. Here we assume that $(\beta_n \cosh \beta_n b - \sinh \beta_n b) \neq 0$ for all $n \in \mathbb{N} \cup \{0\}$. Then expression (8.5) is nothing else but the Fourier series of the function g with respect to the system of eigenfunctions $\sin \beta_n x$. Hence, we obtain formulas for the remaining unknown coefficients A_n:

$$(8.6) \qquad \boxed{A_n = \frac{2}{a} (\beta_n \cosh \beta_n b - \sinh \beta_n b)^{-1} \int_0^a g(x) \sin \beta_n x \, dx.}$$

If $(\beta_n \cosh \beta_n b - \sinh \beta_n b) = 0$ for some $n \in \mathbb{N} \cup \{0\}$, then in general the boundary condition $g(x)$ cannot be expressed as in (8.5). In that case, problem (8.1) can have either no solution, or infinitely many solutions depending on the relation between $g(x)$ and the other data. This means that (8.1) is an ill-posed problem.

Remark 8.1 (Nonhomogeneous Boundary Conditions.) Let us consider again the Laplace equation on a rectangle, but this time let all four boundary conditions be *nonhomogeneous*. (It does not matter which types of boundary conditions (Dirichlet, Neumann, or Robin) are given on particular sides.) The previous example has shown the advantage of the situation when only one boundary condition is nonhomogeneous and all the other three conditions are homogeneous. Using the linearity of the problem, we can decompose the totally nonhomogeneous problem into four partially homogeneous problems which are easy to solve. Schematically, we illustrate the decomposition in Figure 8.2.

8.2 Laplace Equation on Disc

A much more interesting but classical example deals with the *Dirichlet problem for the Laplace equation on a disc*. The rotational invariance of the Laplace operator Δ indicates that the disc is a natural shape for harmonic functions in the plane. So, let us consider the problem

$$(8.7) \qquad \boxed{\begin{aligned} u_{xx} + u_{yy} &= 0 & \text{for } x^2 + y^2 < a^2, \\ u &= h(\theta) & \text{for } x^2 + y^2 = a^2. \end{aligned}}$$

We solve the equation on the disc D with a center at the origin and with the radius a. The boundary condition $h(\theta)$ is given on the circle ∂D. Note that θ is the polar

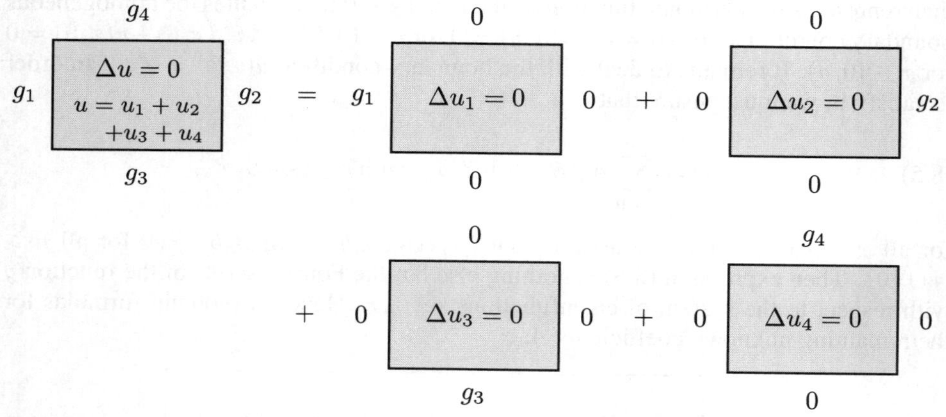

Figure 8.2 *Decomposition of the nonhomogeneous boundary value problem for the Laplace equation on a rectangle.*

coordinate denoting the central angle which is formed by the radius vector of the point (x, y) and the positive half-axis x.

We again use the method of separation of variables, but this time in polar coordinates: $u = R(r)\Theta(\theta)$. If we use the transformation formula $\Delta u = u_{rr} + \frac{1}{r}u_r + \frac{1}{r^2}u_{\theta\theta}$, we rewrite the equation into the form

$$0 = u_{xx} + u_{yy} = u_{rr} + \frac{1}{r}u_r + \frac{1}{r^2}u_{\theta\theta} = R''\Theta + \frac{1}{r}R'\Theta + \frac{1}{r^2}R\Theta''.$$

Dividing by $R\Theta$ (under the assumption $R\Theta \neq 0$) and multiplying by r^2, we obtain two equations

(8.8) $$\Theta'' + \lambda\Theta = 0,$$
(8.9) $$r^2 R'' + rR' - \lambda R = 0.$$

Both these ODEs are easily solvable. We only have to add the appropriate boundary conditions.

For $\Theta(\theta)$, it is natural to introduce periodic boundary conditions:

$$\Theta(\theta + 2\pi) = \Theta(\theta) \quad \text{for } \theta \in \mathbb{R}$$

(that is, $\Theta(0) = \Theta(2\pi)$, $\Theta'(0) = \Theta'(2\pi)$). Hence, we obtain

(8.10) $\quad \lambda_n = n^2 \quad$ and $\quad \Theta(\theta) = A\cos n\theta + B\sin n\theta, \quad n = 0, 1, 2, 3, \ldots.$

The equation for the function R is of Euler type and its solution must be in the form $R(r) = r^\alpha$. Since $\lambda = n^2$, the corresponding characteristic equation is

$$\alpha(\alpha - 1)r^\alpha + \alpha r^\alpha - n^2 r^\alpha = 0,$$

Section 8.3 Poisson Formula

and thus $\alpha = \pm n$. For $n = 1, 2, 3, \ldots$ we obtain $R(r) = \tilde{A}r^n + \tilde{B}r^{-n}$ and the solution u can be written as

$$(8.11) \qquad u = \left(\tilde{A}r^n + \frac{\tilde{B}}{r^n}\right)(A\cos n\theta + B\sin n\theta).$$

For $n = 0$, the functions $R = 1$ and $R = \ln r$ form the couple of linearly independent solutions of equation (8.9). The corresponding u thus assumes the form

$$(8.12) \qquad u = \tilde{A} + \tilde{B}\ln r.$$

For physical reasons, functions (8.11) and (8.12) must be bounded on the whole disc D (this means also at the origin $r = 0$) and thus, in both the cases, we have $\tilde{B} = 0$. We sum up the remaining solutions and write the resulting function u as an infinite series

$$(8.13) \qquad \boxed{u = \frac{1}{2}A_0 + \sum_{n=1}^{\infty} r^n (A_n \cos n\theta + B_n \sin n\theta).}$$

Now, we take into account the nonhomogeneous boundary condition on the boundary $r = a$. It is fulfilled provided the function h is expandable into the Fourier series

$$h(\theta) = \frac{1}{2}A_0 + \sum_{n=1}^{\infty} a^n (A_n \cos n\theta + B_n \sin n\theta).$$

Hence, we easily derive

$$(8.14) \qquad A_n = \frac{1}{\pi a^n} \int_0^{2\pi} h(\phi) \cos n\phi \, d\phi,$$

$$(8.15) \qquad B_n = \frac{1}{\pi a^n} \int_0^{2\pi} h(\phi) \sin n\phi \, d\phi.$$

8.3 Poisson Formula

The previous example has an interesting consequence: the sum of the series (8.13) can be expressed explicitly by an integral formula. If we put (8.14) and (8.15) into (8.13), we obtain

$$\begin{aligned} u(r, \theta) &= \frac{1}{2\pi} \int_0^{2\pi} h(\phi) \, d\phi + \sum_{n=1}^{\infty} \frac{r^n}{\pi a^n} \int_0^{2\pi} h(\phi)[\cos n\phi \cos n\theta + \sin n\phi \sin n\theta] \, d\phi \\ &= \frac{1}{2\pi} \int_0^{2\pi} h(\phi) \left[1 + 2\sum_{n=1}^{\infty} \left(\frac{r}{a}\right)^n \cos n(\theta - \phi)\right] d\phi. \end{aligned}$$

If we express the cosine function using the complex exponential (that is, $\cos t = \frac{1}{2}(e^{it} + e^{-it})$), we can rewrite the expression in the brackets in the following way:

$$1 + 2\sum_{n=1}^{\infty} \left(\frac{r}{a}\right)^n \cos n(\theta - \phi) = 1 + \sum_{n=1}^{\infty} \left(\frac{r}{a}\right)^n e^{in(\theta-\phi)} + \sum_{n=1}^{\infty} \left(\frac{r}{a}\right)^n e^{-in(\theta-\phi)}.$$

The series in this formulation are geometric series with quotients $q = \frac{r}{a} e^{\pm i(\theta-\phi)}$ that, for $r < a$, satisfy the condition $|q| < 1$. Thus, we obtain

$$1 + 2\sum_{n=1}^{\infty} \left(\frac{r}{a}\right)^n \cos n(\theta - \phi) = 1 + \frac{re^{i(\theta-\phi)}}{a - re^{i(\theta-\phi)}} + \frac{re^{-i(\theta-\phi)}}{a - re^{-i(\theta-\phi)}}$$

$$= \frac{a^2 - r^2}{a^2 - 2ar\cos(\theta - \phi) + r^2}.$$

Hence, substituting back into the integral, we arrive at the solution of the original problem (8.7) in the form

(8.16)
$$u(r,\theta) = \frac{a^2 - r^2}{2\pi} \int_0^{2\pi} \frac{h(\phi)}{a^2 - 2ar\cos(\theta - \phi) + r^2}\, d\phi,$$

which is called the *Poisson formula* in polar coordinates. The expression (8.16) implies that the harmonic function inside the circle can be described by its boundary values only.

Now, we go back to the Cartesian coordinates. We denote by $x = (x,y)$ a point inside the circle with polar coordinates (r, θ), and by x' a point on the boundary with polar coordinates (a, ϕ). Then $r = |x|$, $a = |x'|$ and for $|x - x'|$ we have

$$|x - x'|^2 = a^2 + r^2 - 2ar\cos(\theta - \phi).$$

(The reader is asked to draw a picture.) An element of the arc length is in this case $ds = a\, d\phi$. The Poisson formula can be then rewritten into the form

(8.17)
$$u(x) = \frac{a^2 - |x|^2}{2\pi a} \int_{|x'|=a} \frac{u(x')}{|x - x'|^2}\, ds$$

for $x \in D$. Here $u(x') = h(\phi)$ and the integral is considered with respect to the arc length over the whole circumference.

The above calculations are summarized in the following assertion.

Theorem 8.2 *Let $h(\phi)$ be a continuous function on a circle ∂D. Then the Poisson formula (8.16) (or (8.17)) describes the unique harmonic function on the disc D with the property*

$$\lim_{x \to x'} u(x) = h(\phi),$$

where ϕ is the angle corresponding to the point $x' \in \partial D$.

Some important consequences of the Poisson Formula are summarized in Section 10.7.

8.4 Exercises

In the following exercises, r and θ denote the polar coordinates.

1. Solve the equation $u_{xx} + u_{yy} = 1$ in the disc $\{r < a\}$ with the boundary condition $u(x, y) = 0$ on the boundary $r = a$.

   $$[u(r) = \tfrac{1}{4}(r^2 - a^2)]$$

2. Solve the equation $u_{xx} + u_{yy} = 1$ in the annulus $\{a < r < b\}$ with the boundary condition $u(x, y) = 0$ on both boundary circles $r = a$, $r = b$.

   $$[u(r) = \tfrac{r^2}{4} + \tfrac{b^2 - a^2}{4 \ln \frac{a}{b}} \ln r - \tfrac{b^2 \ln a - a^2 \ln b}{4 \ln \frac{a}{b}}]$$

3. Solve the equation $u_{xx} + u_{yy} = 0$ in the rectangle $\{0 < x < a,\ 0 < y < b\}$ with the boundary conditions

   $$u_x(0, y) = -a,\quad u_x(a, y) = 0,$$
   $$u_y(x, 0) = b,\quad u_y(x, b) = 0.$$

   Search for the solution in the form of a quadratic polynomial in x and y.

   $$[u(x, y) = \tfrac{1}{2}x^2 - \tfrac{1}{2}y^2 - ax + by + c,\text{ where } c \text{ is an arbitrary constant}]$$

4. Find a harmonic function $u(x, y)$ in the square $D = \{0 < x < \pi, 0 < y < \pi\}$ with the boundary conditions

   $$u_y(x, 0) = u_y(x, \pi) = 0,$$
   $$u(0, y) = 0,\quad u(\pi, y) = \cos^2 y = \tfrac{1}{2}(1 + \cos 2y).$$

5. Find a harmonic function $u(x, y)$ in the square $D = \{0 < x < 1, 0 < y < 1\}$ with the boundary conditions

   $$u(x, 0) = x,\quad u(x, 1) = 0,$$
   $$u_x(0, y) = 0,\quad u_x(1, y) = y^2.$$

6. Let u be a harmonic function in the disc $D = \{r < 2\}$ and $u = 3 \sin 2\theta + 1$ for $r = 2$. Without finding the concrete form of the solution, determine the value of u at the origin.

   $$[u(0, 0) = 1]$$

7. Solve the equation $u_{xx}+u_{yy} = 0$ in the disc $\{r < a\}$ with the boundary condition $u = 1 + 3\sin\theta$ for $r = a$.

$$[u(r,\theta) = 1 + 3\tfrac{r}{a}\sin\theta]$$

8. Solve the equation $u_{xx} + u_{yy} = 0$ in the disc $\{r < a\}$ with the boundary condition $u = \sin^3\theta$ for $r = a$. Here, use the identity $\sin^3\theta = 3\sin\theta - 4\sin 3\theta$.

$$[u(r,\theta) = 3\tfrac{r}{a}\sin\theta - 4(\tfrac{r}{a})^3 \sin 3\theta]$$

9. Solve the equation $u_{xx} + u_{yy} = 0$ in the domain $\{r > a\}$ (that is, in the *exterior of the disc*) with the boundary condition $u = 1 + 3\sin\theta$ on the boundary $r = a$ and with the condition that the solution u is bounded for $r \to \infty$.

$$[u(r,\theta) = 1 + 3\tfrac{a}{r}\sin\theta]$$

10. Solve the equation $u_{xx}+u_{yy} = 0$ in the disc $\{r < a\}$ with the boundary condition

$$\frac{\partial u}{\partial r} - hu = f(\theta),$$

where $f(\theta)$ is an arbitrary function. Write the solution using the Fourier coefficients of the function f.

11. Derive the Poisson formula for the exterior of the disc in $\mathbb{R}^2$.

12. Find the steady-state temperature distribution inside the annulus $\{1 < r < 2\}$ the outer edge of which ($r = 2$) is heat insulated and the inner edge ($r = 1$) is kept at the temperature described by $\sin^2\theta$.

$$[u(r,\theta) = \tfrac{1}{2}(1 - \tfrac{\ln r}{\ln 2}) + (\tfrac{r^2}{30} - \tfrac{8}{15r^2})\cos 2\theta]$$

13. Find a harmonic function u in the semi-disc $\{r < 1, 0 < \theta < \pi\}$ satisfying the conditions

$$u(r,0) = u(r,\pi) = 0, \quad u(1,\theta) = \pi\sin\theta - \sin 2\theta.$$

14. Solve the equation $u_{xx} + u_{yy} = 0$ in the disc sector $r < a$, $0 < \theta < \beta$ with the boundary conditions

$$u(a,\theta) = 0, \quad u(r,0) = 0, \quad u(r,\beta) = \beta.$$

Search for a function independent of r.

15. Solve the equation $u_{xx} + u_{yy} = 0$ in the quarter-disc $\{x^2 + y^2 < a^2, x > 0, y > 0\}$ with the boundary conditions

$$u(0,y) = u(x,0) = 0, \quad \frac{\partial u}{\partial r} = 1 \text{ for } r = a.$$

Find the solution in the form of an infinite series and write the first two nonzero terms explicitly.

$$[\text{first two terms: } \tfrac{r^2}{2a}\sin 2\theta + \tfrac{r^4}{4a^3}\sin 4\theta]$$

16. Solve the equation $u_{xx} + u_{yy} = 0$ in the domain $\{\alpha < \theta < \beta,\ a < r < b\}$ with the boundary conditions $u = 0$ on both sides $\theta = \alpha$ and $\theta = \beta$, $u = g(\theta)$ on the arc $r = a$, and $u = h(\theta)$ on the arc $r = b$.

17. Solve the boundary value problem for the Laplace equation in the square $K = \{(x, y);\ 0 < x < \pi,\ 0 < y < \pi\}$ for the following data:

    (a) $u_y(x, 0) = u_y(x, \pi) = u_x(0, y) = 0,\quad u_x(\pi, y) = \cos 3y$;

    $$[u(x, y) = \frac{\cosh 3x}{3 \sinh 3\pi} \cos 3y]$$

    (b) $u(0, y) = u_y(x, 0) + u(x, 0) = u_x(\pi, y) = 0,\quad u_x(x, \pi) = \sin \frac{3x}{2}$.

    $$[u(x, y) = \frac{3\cosh(3y/2) - 2\sinh(3y/2)}{3\cosh(3\pi/2) - 2\sinh(3\pi/2)} \sin \frac{3x}{2}]$$

18. Solve the Dirichlet problem
    $$\begin{cases} u_{xx} + u_{yy} = 0 & \text{in } x^2 + y^2 < 1, \\ u(x, y) = x^4 - y^3 & \text{on } x^2 + y^2 = 1. \end{cases}$$

    $$[u(r, \theta) = \tfrac{3}{8} - \tfrac{3}{4} r \sin\theta + \tfrac{r^2}{2}\cos 2\theta + \tfrac{r^3}{4}\sin 3\theta + \tfrac{r^4}{8}\cos 4\theta]$$

19. Let $U_n(x, y)$ be the solution of the problem
    $$\begin{cases} u_{xx} + u_{yy} = 0 & \text{in } x^2 + y^2 < 1, \\ u(x, y) = y^n & \text{on } x^2 + y^2 = 1. \end{cases}$$

    Prove:

    (a)
    $$U_{2m}(r, \theta) = \frac{1}{2^{2m}} \left(2 \sum_{k=0}^{m-1} (-1)^{m+k} \binom{2m}{k} r^{2(m-k)} \cos(2(m-k)\theta) + \binom{2m}{m} \right),$$

    $$U_{2m+1}(r, \theta) = \frac{1}{2^{2m}} \left(\sum_{k=0}^{m-1} (-1)^{m+k} \binom{2m+1}{k} r^{2(m-k)+1} \sin((2(m-k)+1)\theta) \right).$$

    (b)
    $$1 \geq U_2(x, y) \geq U_4(x, y) \geq \cdots \geq U_{2m}(x, y) \geq 0,$$
    $$1 \geq U_1(x, y) \geq U_3(x, y) \geq \cdots \geq U_{2m+1}(x, y) \geq 0 \quad \text{if } y \geq 0,$$
    $$-1 \leq U_1(x, y) \leq U_3(x, y) \leq \cdots \leq U_{2m+1}(x, y) \leq 0 \quad \text{if } y \leq 0.$$

    (c) The series $\sum_{m=1}^{\infty} U_{2m}(x, y)$ is divergent for $x^2 + y^2 < 1$, while the series $\sum_{m=1}^{\infty} q^m U_{2m+1}(x, y)$ is convergent for $0 \leq q < 1$, $x^2 + y^2 < 1$.

20. Solve the Poisson equation $u_{xx} + u_{yy} = f(x,y)$ in the unit square ($0 < x < 1$, $0 < y < 1$) for the following data.

(a) $f(x,y) = x$, $u(x,0) = u(x,1) = u(0,y) = u(1,y) = 0$.

$$[u(x,y) = \tfrac{8}{\pi^4} \sum_{k=0}^{\infty} \sum_{m=1}^{\infty} \tfrac{(-1)^m}{(m^2+(2k+1)^2)m(2k+1)} \sin m\pi x \sin(2k+1)\pi y]$$

(b) $f(x,y) = \sin \pi x$, $u(x,0) = u(0,y) = u(1,y) = 0$, $u(x,1) = x$.

$$[u(x,y) = \tfrac{-4 \sin \pi x}{\pi^3} \sum_{k=0}^{\infty} \tfrac{\sin(2k+1)\pi y}{(1+(2k+1)^2)(2k+1)} + \tfrac{2}{\pi} \sum_{n=1}^{\infty} \tfrac{(-1)^{n+1} \sin n\pi x \sinh n\pi y}{n \sinh n\pi}]$$

(c) $f(x,y) = xy$, $u(x,0) = u(0,y) = u(1,y) = 0$, $u(x,1) = x$.

21. Solve the equation $u_{xx} + u_{yy} = 3u - 1$ inside the unit square ($0 < x < 1$, $0 < y < 1$) with u vanishing on the boundary.

$$[u(x,y) = \tfrac{16}{\pi^2} \sum_{l=0}^{\infty} \sum_{k=0}^{\infty} \tfrac{\sin(2k+1)\pi x \sin(2l+1)\pi y}{(2l+1)(2k+1)(3+\pi^2((2l+1)^2+(2k+1)^2))}]$$

Exercises 1–16 are taken from Strauss [19], 17–19 from Stavroulakis and Tersian [18], and 20–21 from Asmar [3].

9 Methods of Integral Transforms

In this chapter we introduce another class of methods that can be used for solving the initial value or initial boundary value problems for evolution equations. These are the so called *methods of integral transforms*. The fundamental ones are the Laplace and the Fourier transforms.

9.1 Laplace Transform

The reader could probably meet with the Laplace transform when solving linear ODEs with constant coefficients, where the equations are transformed in this way to algebraic equations. This idea can be easily extended to PDEs, where the transformation leads to the decrease of the number of independent variables. PDEs in two variables are thus reduced to ODEs.

Let $u = u(t)$ be a *piecewise continuous* function on $[0, \infty)$ that "does not grow too fast". Let us assume, for example, that u is of *exponential order*, which means that $|u(t)| \leq ce^{at}$ for t large enough, where $a, c > 0$ are appropriate constants. The Laplace transform of the function u is then defined by the formula

(9.1) $$(\mathcal{L}u)(s) \equiv U(s) = \int_0^\infty u(t)e^{-st}\, dt.$$

Here U and s are the *transformed variables*, U is the dependent one, s is the independent one, and U is defined for $s > a$ with $a > 0$ depending on $u(t)$. The function U is called the *Laplace image* of the function u, which is then called the *original*. The Laplace transform is a linear mapping, that is,

$$\mathcal{L}(c_1 u + c_2 v) = c_1 \mathcal{L}u + c_2 \mathcal{L}v,$$

where c_1, c_2 are arbitrary constants. If we know the Laplace image $U(s)$, then the original $u(t)$ can be obtained by the *inverse Laplace transform* of the image $U(s)$: $\mathcal{L}^{-1}U = u$. The couples of Laplace images and their originals can be found in tables, or, in some cases, the transformation can be done using various software packages.

An important property of the Laplace transform, as well as of other integral transforms, is the fact that it turns differential operators in originals into multiplication operators in images. The following formulas hold:

(9.2) $(\mathcal{L}u')(s) = sU(s) - u(0),$

(9.3) $(\mathcal{L}u^{(n)})(s) = s^n U(s) - s^{n-1}u(0) - s^{n-2}u'(0) - \cdots - u^{(n-1)}(0),$

if the derivatives considered are transformable (i.e., piecewise continuous functions of exponential order). To be precise, we should write $\lim_{t \to 0_+} u(t)$, $\lim_{t \to 0_+} u'(t), \ldots$

instead of $u(0), u'(0), \dots$. However, without loss of generality, we can assume that the function u and its derivatives are continuous from the right at 0. Relations (9.2), (9.3) can be easily derived directly from the definition using integration by parts (the reader is asked to do it in detail.). Applying the Laplace transform to a linear ODE with constant coefficients, we obtain a linear algebraic equation for the unknown function $U(s)$. After solving it, we find the original function $u(t)$ by the inverse transform.

The same idea can be exploited also when solving PDEs for functions of two variables, say $u = u(x,t)$. The transformation will be done with respect to the time variable $t \geq 0$, the spatial variable x will be treated as a parameter unaffected by this transform. In particular, we define the Laplace transform of a function $u(x,t)$ by the formula

(9.4) $$\boxed{(\mathcal{L}u)(x,s) \equiv U(x,s) = \int_0^\infty u(x,t)\mathrm{e}^{-st}\,\mathrm{d}t.}$$

The time derivatives are transformed in the same way as in the case of functions of one variable, that is, for example,

$$(\mathcal{L}u_t)(x,s) = sU(x,s) - u(x,0).$$

The spatial derivatives remain unchanged, that is,

$$(\mathcal{L}u_x)(x,s) = \int_0^\infty \frac{\partial}{\partial x}u(x,t)\mathrm{e}^{-st}\,\mathrm{d}t = \frac{\partial}{\partial x}\int_0^\infty u(x,t)\mathrm{e}^{-st}\,\mathrm{d}t = U_x(x,s).$$

Thus, applying the Laplace transform to a PDE in two variables x and t, we obtain an ODE in the variable x and with the parameter s.

Example 9.1 (Constant Boundary Condition.) Using the Laplace transform, we solve the following initial boundary value problem for the diffusion equation. Let $u = u(x,t)$ denote the concentration of a chemical contaminant dissolved in a liquid on a half-infinite domain $x > 0$. Let us assume that, at time $t = 0$, the concentration is zero. On the boundary $x = 0$, constant unit concentration of the contaminant is kept for $t > 0$. Assuming the unit diffusion constant, the behavior of the system is described by a mathematical model

(9.5) $$\begin{cases} u_t - u_{xx} = 0, & x > 0,\ t > 0, \\ u(x,0) = 0, \\ u(0,t) = 1, & u(x,t) \text{ bounded.} \end{cases}$$

Here the boundedness assumption is related to the physical properties of the model and its solution. If we apply the Laplace transform to both sides of the equation, we obtain the following relation for the image U:

$$sU(x,s) - U_{xx}(x,s) = 0.$$

This is an ODE with respect to the variable x and with real positive parameter s. Its general solution has the form

$$U(x,s) = a(s)\mathrm{e}^{-\sqrt{s}x} + b(s)\mathrm{e}^{\sqrt{s}x}.$$

Section 9.1 Laplace Transform

Since we require the solution u to be bounded in both variables x and t, the image U must be bounded in x as well. Thus, $b(s)$ must vanish, and hence

$$U(x,s) = a(s)e^{-\sqrt{s}x}.$$

Now, we apply the Laplace transform to the boundary condition obtaining $U(0,s) = \mathcal{L}(1) = 1/s$. It implies $a(s) = 1/s$ and the transformed solution has the form

$$U(x,s) = \frac{1}{s}e^{-\sqrt{s}x}.$$

Using the tables of the Laplace transform or some of the software packages, we easily find out that

$$u(x,t) = \operatorname{erfc}\left(\frac{x}{\sqrt{4t}}\right),$$

where erfc is the function defined by the relation

$$\operatorname{erfc}(y) = 1 - \frac{2}{\sqrt{\pi}} \int_0^y e^{-r^2}\,dr = 1 - \operatorname{erf}(y).$$

$\square$

In the previous example, we were able to find the original $u(x,t)$ to the Laplace image $U(x,s)$ using tables or software packages. There exists a general formula for inverse Laplace transform, which is based on theory of functions of complex variables (see, e.g., [21]). It has a theoretical character, and from the practical point of view, it is used very rarely. In most cases, it is more or less useless.

In some cases, instead of above mentioned inverse formula, we can exploit another useful tool, which is stated in the so called *Convolution Theorem*.

Theorem 9.2 *Let u and v be piecewise continuous functions on the interval $(0, \infty)$, both of exponential order, and let $U = \mathcal{L}u$, $V = \mathcal{L}v$ be their Laplace images. Let us denote by*

$$(u * v)(t) = \int_0^t u(t-\tau)v(\tau)\,d\tau$$

the convolution of functions u and v (which is also of exponential order). Then

$$\mathcal{L}(u * v)(s) = (\mathcal{L}u)(s)\,(\mathcal{L}v)(s) = U(s)V(s).$$

Remark 9.3 In particular, it follows from Theorem 9.2 that

$$u * v = \mathcal{L}^{-1}(\mathcal{L}u\,\mathcal{L}v).$$

Notice that the Laplace transform is additive, however, it is not multiplicative!

Example 9.4 (Non-Constant Boundary Condition.) Let us consider the same situation as in the previous example with the only change—the boundary condition will be a time-dependent function:

(9.6)
$$\begin{cases} u_t - u_{xx} = 0, & x > 0,\ t > 0, \\ u(x,0) = 0, \\ u(0,t) = f(t), & u(x,t)\ \text{bounded}. \end{cases}$$

Applying the Laplace transform to the equation, we obtain again the ODE

$$sU(x,s) - U_{xx}(x,s) = 0,$$

the solution of which has the form

$$U(x,s) = a(s)e^{-\sqrt{s}x}.$$

Here we have used the boundedness assumption. The transformation of the boundary condition (we assume that the Laplace transform of f does exist) leads to the relation $U(0,s) = F(s)$ with $F = \mathcal{L}f$. Hence $a(s) = F(s)$ and the solution in images takes the form

$$U(x,s) = F(s)e^{-\sqrt{s}x}.$$

The Convolution Theorem and Remark 9.3 now imply

$$u = \mathcal{L}^{-1}U = \mathcal{L}^{-1}F * \mathcal{L}^{-1}(e^{-\sqrt{s}x}) = f * \mathcal{L}^{-1}(e^{-\sqrt{s}x}).$$

If we exploit the knowledge of the transform relation

$$\mathcal{L}^{-1}\left(e^{-\sqrt{s}x}\right) = \frac{x}{\sqrt{4\pi t^3}}e^{-x^2/4t},$$

we obtain the solution of the original problem in the form

$$u(x,t) = \int_0^t \frac{x}{\sqrt{4\pi(t-\tau)^3}} e^{-x^2/4(t-\tau)} f(\tau)\,d\tau.$$

$\square$

Example 9.5 (Forced Vibrations of "Half-Infinite String".) Let us consider a "half-infinite string" that has one end fixed at the origin and that lies motionless at time $t = 0$. The string is set in motion by acting of a force $f(t)$. The string behavior is then modeled by the problem

(9.7)
$$\begin{cases} u_{tt} - c^2 u_{xx} = f(t), & x > 0,\ t > 0, \\ u(0,t) = 0, \\ u(x,0) = u_t(x,0) = 0, & u(x,t)\ \text{bounded}. \end{cases}$$

Section 9.1 Laplace Transform

We transform both sides of the equation with respect to the time variable and use the initial condition. Thus we obtain

$$-c^2 U_{xx}(x,s) + s^2 U(x,s) = F(s).$$

This is an ODE in the x-variable with constant coefficients and a non-zero right-hand side. Its solution is the sum of the solution U_H of the homogeneous equation: $U_H(x,s) = A(s)e^{-sx/c} + B(s)e^{sx/c}$, and the particular solution U_P of the nonhomogeneous equation: $U_P(x,s) = F(s)/s^2$. Hence

$$U(x,s) = A(s)e^{-\frac{s}{c}x} + B(s)e^{\frac{s}{c}x} + \frac{F(s)}{s^2}.$$

Since we require the original solution $u(x,t)$ to be bounded, the transformed solution $U(x,s)$ must be bounded for $x > 0$, $s > 0$ as well, and thus $B(s) = 0$. The transformed boundary condition implies $A(s) = -F(s)/s^2$ and thus

$$U(x,s) = F(s)\frac{1 - e^{-\frac{s}{c}x}}{s^2}.$$

In the inverse Laplace transform, we use the Convolution Theorem and the relations

$$\mathcal{L}^{-1}(1/s^2) = t, \quad \mathcal{L}^{-1}(\frac{e^{-sx/c}}{s^2}) = (t - \frac{x}{c})\mathcal{H}(t - \frac{x}{c}),$$

where $\mathcal{H}$ is the Heaviside step function, that is, $\mathcal{H}(t) = 0$ for $t \le 0$, $\mathcal{H}(t) = 1$ for $t > 0$. The solution of the original problem then has the form

$$u(x,t) = f(t) * \mathcal{L}^{-1}\left(\frac{1 - e^{-\frac{s}{c}x}}{s^2}\right) = f(t) * \left[t - (t - \frac{x}{c})\mathcal{H}(t - \frac{x}{c})\right]$$

or

$$u(x,t) = \int_0^t f(t - \tau)\left[\tau - (\tau - \frac{x}{c})\mathcal{H}(\tau - \frac{x}{c})\right] d\tau.$$

□

The following example illustrates one particular interesting case of Example 9.5.

Example 9.6 (String Vibrations due to Gravity Acceleration.) If the only acting external force in Example 9.5 is the gravity acceleration g, we solve the wave equation in the form

$$u_{tt} - c^2 u_{xx} = -g.$$

Under the same initial and boundary conditions as in the previous example (that is, $u(x,0) = u_t(x,0) = 0$ for $x > 0$, and $u(0,t) = 0$ for $t > 0$), the solution assumes the form

$$u(x,t) = -g\int_0^t \left[\tau - (\tau - \frac{x}{c})\mathcal{H}(\tau - \frac{x}{c})\right]d\tau$$

$$= -g\left[\frac{1}{2}t^2 - \int_0^t (\tau - \frac{x}{c})\mathcal{H}(\tau - \frac{x}{c})\,d\tau\right].$$

By simple calculation, we express the integral on the right-hand side and obtain the final formulation of the solution:

$$u(x,t) = \begin{cases} -\dfrac{g}{2}(t^2 - (t - \dfrac{x}{c})^2) & \text{for } 0 < x < ct, \\ -\dfrac{gt^2}{2} & \text{for } x > ct. \end{cases}$$

Figure 9.1 shows the solution on several time levels. This example models a half-infinite string with one fixed end, which falls from the zero (horizontal) position due to the gravitation. Recalling that the position of the free-falling mass point is described by the function $-gt^2/2$, we see that, for x bigger than ct, the string falls freely. The remaining part of the string ($x < ct$) falls more slowly due to the fixed end. Notice that this effect propagates from the point $x = 0$ to the right at the speed equal to the constant c (it propagates along the characteristic $x - ct = 0$).

□

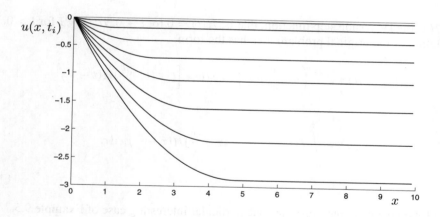

Figure 9.1 *A string falling due to the gravitation.*

9.2 Fourier Transform

The Fourier transform is another integral transform with properties similar to the Laplace transform. Since it again turns differentiation of the originals into multiplication of

Section 9.2 Fourier Transform

the images, it is a useful tool in solving differential equations. Contrary to the Laplace transform, which usually uses the time variable, the Fourier transform is applied to the spatial variable on the whole real line.

First, we start with functions of one spatial variable. The Fourier transform of a function $u = u(x)$, $x \in \mathbb{R}$, is a mapping defined by the formula

(9.8)
$$(\mathcal{F}u)(\xi) \equiv \hat{u}(\xi) = \int_{-\infty}^{\infty} u(x) e^{-i\xi x}\, dx.$$

If $|u|$ is integrable in $\mathbb{R}$, that is, $\int_{-\infty}^{\infty} |u|\, dx < \infty$, then $\hat{u}$ exists. However, the theory of the Fourier transform usually works with a smaller set of functions. We define the so called *Schwartz space* $\mathcal{S}$ as the space of functions on $\mathbb{R}$ that have continuous derivatives of all orders and that, together with their derivatives, decrease to zero for $x \to \pm\infty$ more rapidly than $|x|^{-n}$ for an arbitrary $n \in \mathbb{N}$. It means

$$\mathcal{S} = \{u \in C^{\infty};\ \exists M = M(u) \in \mathbb{R},\ \left|\frac{d^k u}{dx^k}\right| \le M \frac{1}{|x|^n} \text{ for } |x| \to \infty,\ k \in \mathbb{N}\cup\{0\};\ n \in \mathbb{N}\}.$$

It can be shown that, if $u \in \mathcal{S}$, then $\hat{u} \in \mathcal{S}$, and vice versa. We say that the Schwartz space $\mathcal{S}$ is closed with respect to the Fourier transform.

It is important to mention that there exists no established convention how to define the Fourier transform. In literature, we can meet an equivalent of the definition formula (9.8) with the constant $1/\sqrt{2\pi}$ or $1/(2\pi)$ in front of the integral. There also exist definitions with positive sign in the exponent. The reader should keep this fact in mind while working with various sources or using the transformation tables.

The fundamental formula of the Fourier transform is that for the image of the kth derivative $u^{(k)}$:

(9.9)
$$(\mathcal{F}u^{(k)})(\xi) = (i\xi)^k \hat{u}(\xi), \quad u \in \mathcal{S}.$$

The derivation of this formula is based on integration by parts where all "boundary values" vanish due to zero values of the function and its derivatives at infinity. In the case of functions of two variables, say $u = u(x,t)$, the variable t plays the role of a parameter and we define

(9.10)
$$(\mathcal{F}u)(\xi,t) \equiv \hat{u}(\xi,t) = \int_{-\infty}^{\infty} u(x,t) e^{-i\xi x}\, dx.$$

The derivatives with respect to the spatial variable are transformed analogously as (9.9), the derivatives with respect to the time variable t stay unchanged; thus, for instance,

$$\begin{aligned}
(\mathcal{F}u_x)(\xi,t) &= (i\xi)\hat{u}(\xi,t), \\
(\mathcal{F}u_{xx})(\xi,t) &= (i\xi)^2 \hat{u}(\xi,t), \\
(\mathcal{F}u_t)(\xi,t) &= \hat{u}_t(\xi,t).
\end{aligned}$$

The PDE in two variables x, t passes under the Fourier transform to an ODE in the t-variable. By its solving, we obtain the transformed function (the image) $\hat{u}$ which can be converted to the original function u by the *inverse Fourier transform*

(9.11) $$(\mathcal{F}^{-1}\hat{u})(x,t) \equiv u(x,t) = \frac{1}{2\pi} \int_{-\infty}^{\infty} \hat{u}(\xi,t) e^{i\xi x} \, d\xi.$$

In comparison with the inverse Laplace transform, where the general inverse formula is quite complicated, this relation is very simple. Nevertheless, it is convenient to use the transformation tables or some software packages when solving particular problems.

Example 9.7 Let us determine the Fourier transform of the Gaussian function $u(x) = e^{-ax^2}$, $a > 0$, that is,

$$\hat{u}(\xi) = \int_{-\infty}^{\infty} e^{-ax^2} e^{-i\xi x} \, dx.$$

If we differentiate this relation with respect to the ξ-variable and integrate it by parts, we obtain

$$\hat{u}'(\xi) = -i \int_{-\infty}^{\infty} x e^{-ax^2} e^{-i\xi x} \, dx$$

$$= \frac{i}{2a} \int_{-\infty}^{\infty} \frac{d}{dx}(e^{-ax^2}) e^{-i\xi x} \, dx$$

$$= \frac{i^2 \xi}{2a} \int_{-\infty}^{\infty} e^{-ax^2} e^{-i\xi x} \, dx = \frac{-\xi}{2a} \hat{u}(\xi).$$

Thus we have arrived at the differential equation $\hat{u}' = \frac{-\xi}{2a}\hat{u}$, whose general solution has the form

$$\hat{u}(\xi) = C e^{-\xi^2/(4a)}.$$

The constant C can be determined from the relation

$$\hat{u}(0) = \int_{-\infty}^{\infty} e^{-ax^2} \, dx = \sqrt{\frac{\pi}{a}}.$$

Hence, we have

$$\mathcal{F}(e^{-ax^2}) = \sqrt{\frac{\pi}{a}} e^{-\xi^2/(4a)},$$

which means that the Fourier transform of the Gaussian function is the Gaussian function again. The same holds also for the inverse transform. (The reader is asked to give the reasons.)

□

Section 9.2 Fourier Transform

As in the case of the Laplace transform, the Convolution Theorem holds true for the Fourier transform and is directly applicable for solving differential equations. However, the convolution of two functions u and v is now defined in the following way:

$$(u * v)(x) = \int_{-\infty}^{\infty} u(x - y) v(y) \, dy.$$

Theorem 9.8 *If $u, v \in \mathcal{S}$, then*

$$\mathcal{F}(u * v)(\xi) = \hat{u}(\xi) \hat{v}(\xi).$$

Example 9.9 (Cauchy Problem for Diffusion Equation.) Now we use the Fourier transform for derivation of the solution of the Cauchy problem for the diffusion equation which we treated in Chapter 5. Thus, let us consider the problem

(9.12) $$\begin{cases} u_t - k u_{xx} = 0, & x \in \mathbb{R},\ t > 0, \\ u(x, 0) = \varphi(x). \end{cases}$$

Let us assume $\varphi \in \mathcal{S}$. Using the Fourier transform, we reduce the diffusion equation to the form

$$\hat{u}_t = -\xi^2 k \hat{u},$$

which is an ODE in the t-variable for the required function $\hat{u}(\xi, t)$ with the parameter ξ. Its solution is

$$\hat{u}(\xi, t) = C e^{-\xi^2 k t}.$$

The initial condition implies $\hat{u}(\xi, 0) = \hat{\varphi}(\xi)$ and thus $C = \hat{\varphi}(\xi)$. The solution in images then assumes the form

$$\hat{u}(\xi, t) = \hat{\varphi}(\xi) e^{-\xi^2 k t}.$$

If we use the knowledge of the transformation relation

$$\mathcal{F}\left(\frac{1}{\sqrt{4\pi k t}} e^{-x^2/(4kt)}\right) = e^{-\xi^2 k t}$$

and the Convolution Theorem, we obtain the solution of the Cauchy problem (9.12) in the form

(9.13) $$u(x, t) = \int_{-\infty}^{\infty} \frac{1}{\sqrt{4\pi k t}} e^{-(x-y)^2/(4kt)} \varphi(y) \, dy,$$

which is exactly formula (5.11) derived in Chapter 5. □

Remark 9.10 When using the Fourier transform, we have obtained the solution (9.13) under the assumption that the initial condition φ belongs to the Schwartz space. However, once the solution is derived, we can try to show that it exists even under weaker assumptions on the function φ. It can be proved, for instance, that the function u in (9.13) solves problem (9.12) provided φ is a continuous and bounded function on $\mathbb{R}$.

Example 9.11 (**Cauchy Problem for Wave Equation.**) Let us solve the Cauchy problem

(9.14) $$\begin{cases} u_{tt} - c^2 u_{xx} = 0, & x \in \mathbb{R}, \ t > 0, \\ u(x,0) = \varphi(x), & u_t(x,0) = \psi(x). \end{cases}$$

We apply again the Fourier transform with respect to the spatial variable to the equation and both initial conditions. Thus we obtain the transformed problem

$$\begin{cases} \hat{u}_{tt}(\xi,t) + c^2 \xi^2 \hat{u}(\xi,t) = 0, \\ \hat{u}(\xi,0) = \hat{\varphi}(\xi), \quad \hat{u}_t(\xi,0) = \hat{\psi}(\xi). \end{cases}$$

Its solution is the function

$$\hat{u}(\xi,t) = \hat{\varphi}(\xi) \cos c\xi t + \frac{1}{c\xi} \hat{\psi}(\xi) \sin c\xi t.$$

The solution of the original problem is then found by the inverse Fourier transform:

(9.15) $$u(x,t) = \frac{1}{2\pi} \int_{-\infty}^{\infty} [\hat{\varphi}(\xi) \cos c\xi t + \frac{1}{c\xi} \hat{\psi}(\xi) \sin c\xi t] e^{i\xi x} \, d\xi.$$

This integral expression, where, moreover, the Fourier transforms of the initial conditions occur, is not very transparent. Nevertheless, it can be converted to d'Alembert's formula (4.15) derived in Chapter 4. Indeed, substituting the complex representation of the sine and cosine functions into (9.15), we obtain

(9.16) $$u(x,t) = \frac{1}{2\pi} \int_{-\infty}^{\infty} \frac{1}{2} \hat{\varphi}(\xi) \left(e^{ic\xi t} + e^{-ic\xi t} \right) e^{i\xi x} \, d\xi$$

$$+ \frac{1}{2\pi} \int_{-\infty}^{\infty} \frac{1}{2ic\xi} \hat{\psi}(\xi) \left(e^{ic\xi t} - e^{-ic\xi t} \right) e^{i\xi x} \, d\xi.$$

The first integral on the right-hand side can be written as

$$\frac{1}{4\pi} \int_{-\infty}^{\infty} \left(\hat{\varphi}(\xi) e^{i(x+ct)\xi} + \hat{\varphi}(\xi) e^{i(x-ct)\xi} \right) d\xi,$$

Section 9.2 Fourier Transform

which is (using the definition of the inverse Fourier transform (9.11)) exactly the first term in d'Alembert's formula

$$\frac{1}{2}\left(\varphi(x+ct)+\varphi(x-ct)\right).$$

Similarly, the second integral term in (9.16) equals

$$\frac{1}{4\pi c}\int_{-\infty}^{\infty}\frac{1}{i\xi}\hat{\psi}(\xi)\left(e^{i(x+ct)\xi}-e^{i(x-ct)\xi}\right)d\xi=\frac{1}{4\pi c}\int_{-\infty}^{\infty}\hat{\psi}(\xi)\int_{x-ct}^{x+ct}e^{iy\xi}\,dy\,d\xi.$$

Changing the order of integration and using again the inverse Fourier transform, we obtain the second term in d'Alembert's formula

$$\frac{1}{2c}\int_{x-ct}^{x+ct}\psi(y)\,dy.$$

□

Remark 9.12 In some cases, the methods of integral transforms are applicable also to equations with non-constant coefficients. Let us consider, for example, the Cauchy problem for the transport equation

$$\begin{cases} tu_x+u_t=0, & x\in\mathbb{R},\ t>0,\\ u(x,0)=f(x).\end{cases}$$

Since the varying coefficient is—in this case—the time variable t, we use the Fourier transform with time playing the role of a parameter. We have

$$\mathcal{F}(tu_x)=t\mathcal{F}(u_x)=i\xi t\hat{u}.$$

Transforming the equation and the initial conditions, we obtain

$$i\xi t\hat{u}+\hat{u}_t=0,\quad \hat{u}(\xi,0)=\hat{f}(\xi),$$

and hence

$$\hat{u}(\xi,t)=\hat{f}(\xi)e^{-i\frac{t^2}{2}\xi}.$$

By the inverse Fourier transform (e.g., using the transformation formulas), we obtain the solution of the original equation in the form

$$u(x,t)=f\left(x-\frac{t^2}{2}\right).$$

Remark 9.13 (Laplace and Poisson Equations.) The Laplace and Poisson equations can also be solved, in some cases, by the method of integral transforms.

As an example, let us consider the problem

$$\begin{cases} u_{xx} + u_{yy} = 0, & x \in \mathbb{R}, \ y > 0, \\ u(x,0) = f(x), \\ u(x,y) \text{ bounded for } y \to \infty. \end{cases}$$

We will search for a solution using the Fourier transform with respect to x. Its application to our problem leads to the equation

$$\hat{u}_{yy} - \xi^2 \hat{u} = 0,$$

whose general solution is $\hat{u}(\xi, y) = a(\xi)e^{-\xi y} + b(\xi)e^{\xi y}$. The boundedness assumption implies

$$b(\xi) = 0 \quad \text{for } \xi > 0,$$
$$a(\xi) = 0 \quad \text{for } \xi < 0.$$

Hence, $\hat{u}(\xi, y) = c(\xi)e^{-|\xi|y}$. Here, a, b, c are arbitrary functions. If we take into account the boundary condition, we derive $c(\xi) = \hat{f}(\xi)$ and thus

$$\hat{u}(\xi, y) = e^{-|\xi|y}\hat{f}(\xi).$$

The inverse transformation leads to the solution of the original problem in the form of a convolution:

$$u(x,y) = \left(\frac{y}{\pi}\frac{1}{x^2 + y^2}\right) * f = \frac{y}{\pi}\int_{-\infty}^{\infty}\frac{f(\tau)\,d\tau}{(x-\tau)^2 + y^2}.$$

The reader should notice that in this convolution y is just a parameter.

9.3 Exercises

1. Derive the following transform relations:

   (a) $\mathcal{L}\{1\} = \dfrac{1}{s}, \quad s > 0,$

   (b) $\mathcal{L}\{t\} = \dfrac{1}{s^2},$

   (c) $\mathcal{L}\{t^n\} = \dfrac{n!}{s^n}, \quad n \in \mathbb{N}, \ s > 0,$

   (d) $\mathcal{L}\{e^{at}\} = \dfrac{1}{s-a}, \quad s > a,$

   (e) $\mathcal{L}\{\sin(at)\} = \dfrac{a}{s^2+a^2}, \quad s > 0,$

   (f) $\mathcal{L}\{\cos(at)\} = \dfrac{s}{s^2+a^2}, \quad s > 0.$

Section 9.3 Exercises

2. Derive the following basic properties of the Laplace transform (here $U = \mathcal{L}\{u\}$):

   (a) $\mathcal{L}\{t^n u(t)\} = (-1)^n U^{(n)}(s)$,

   (b) $\mathcal{L}\{e^{at} u(t)\} = U(s-a)$,

   (c) $\mathcal{L}\{\int_0^t u(\tau) d\tau\} = \frac{1}{s} U(s), \quad s > 0$,

   (d) $\mathcal{L}\{\frac{1}{t} u(t)\} = \int_s^\infty U(\sigma) d\sigma$,

   (e) $\mathcal{L}\{u(ct)\} = \frac{1}{c} U(\frac{s}{c}), \quad c > 0$.

3. Using substitution and Fubini's Theorem, prove the formulas in Theorems 9.2 and 9.8.

4. Using the Laplace transform method, solve the following initial boundary value problems. Simplify the results as much as possible.

   (a)
   $$\begin{cases} u_t = u_{xx}, & x > 0, \ t > 0, \\ u(0,t) = 70, \\ u(x,0) = 0. \end{cases}$$

   $[u(x,t) = 70\,\text{erfc}\,(\frac{x}{\sqrt{4t}})]$

   (b)
   $$\begin{cases} u_{tt} = u_{xx} + t, & x > 0, \ t > 0, \\ u(0,t) = 0, \\ u(x,0) = 0, \ u_t(x,0) = 0. \end{cases}$$

   $[u(x,t) = \frac{1}{3!} t^3 - \frac{1}{3!} \mathcal{H}(t-x)(t-x)^3]$

   (c)
   $$\begin{cases} u_{tt} = u_{xx} + e^{-t}, & x > 0, \ t > 0, \\ u(0,t) = 0, \\ u(x,0) = 0, \ u_t(x,0) = 0. \end{cases}$$

   (d)
   $$\begin{cases} u_{tt} = u_{xx} - g, & x > 0, \ t > 0, \\ u(0,t) = 0, \\ u(x,0) = 0, \ u_t(x,0) = 1. \end{cases}$$

   $[u(x,t) = t - (t-x)\mathcal{H}(t-x) - \frac{g}{2}(t^2 - (t-x)^2 \mathcal{H}(t-x))]$

   (e)
   $$\begin{cases} u_{tt} = u_{xx} + t^2, & x > 0, \ t > 0, \\ u(0,t) = 0, \\ u(x,0) = 0, \ u_t(x,0) = 0. \end{cases}$$

(f)
$$\begin{cases} u_{tt} = u_{xx}, & x > 0,\ t > 0, \\ u(0,t) = \sin t, \\ u(x,0) = 0,\ u_t(x,0) = 1. \end{cases}$$

$$[u(x,t) = t + \sin(t-x)\mathcal{H}(t-x) - (t-x)\mathcal{H}(t-x)]$$

(g)
$$\begin{cases} u_{tt} = u_{xx}, & x > 0,\ t > 0, \\ u(0,t) = 0, \\ u(x,0) = 0,\ u_t(x,0) = 1. \end{cases}$$

5. Using the Laplace transform method, solve the initial boundary value problem
$$\begin{cases} u_{tt} = u_{xx} + \sin \pi x, & 0 < x < 1,\ t > 0, \\ u(0,t) = 0,\ u(1,t) = 0, \\ u(x,0) = 0,\ u_t(x,0) = 0. \end{cases}$$

6. Show that the solution of the initial boundary value problem
$$\begin{cases} u_t = k u_{xx}, & x > 0,\ t > 0, \\ u(0,t) = T_0, \\ u(x,0) = T_1 \end{cases}$$

is given by

$$u(x,t) = (T_0 - T_1)\operatorname{erfc}\left(\frac{x}{\sqrt{4kt}}\right) + T_1 = (T_0 - T_1)\operatorname{erf}\left(\frac{x}{\sqrt{4kt}}\right) + T_0.$$

7. Using the Laplace transform, solve the initial boundary value problem
$$\begin{cases} u_t = k u_{xx}, & x > 0,\ t > 0, \\ u(0,t) = \alpha,\ \lim_{x \to \infty} u(x,t) = \beta, \\ u(x,0) = \gamma. \end{cases}$$

8. Using the Laplace transform, solve the initial boundary value problem
$$\begin{cases} u_{tt} = c^2 u_{xx}, & x > 0,\ t > 0, \\ u_x(0,t) = \alpha,\ \lim_{x \to \infty} u(x,t) = \beta, \\ u(x,0) = \varphi(x),\ u_t(x,0) = \psi(x). \end{cases}$$

9. Use the Laplace transform to solve the problem
$$\begin{cases} u_{tt} = c^2 u_{xx} + \cos \omega t \sin \pi x, & 0 < x < 1,\ t > 0, \\ u(0,t) = u(1,t) = 0, \\ u(x,0) = u_t(x,0) = 0. \end{cases}$$

Assume that $\omega > 0$ and be careful of the case $\omega = c\pi$. Check your answer by direct differentiation.

Section 9.3   Exercises

10. Prove the following transform relations (here $\hat{u}(\xi) = \mathcal{F}\{u(x)\}$):

    (a) $u(x) = \begin{cases} 1, & |x| < a, \\ 0, & |x| > a, \end{cases} \quad \hat{u}(\xi) = 2\dfrac{\sin a\xi}{\xi}$,

    (b) $u(x) = \begin{cases} 1 - \frac{|x|}{a}, & |x| < a, \\ 0, & |x| > a, \end{cases} \quad \hat{u}(\xi) = 4\dfrac{\sin^2(a\xi/2)}{a\xi^2}$,

    (c) $u(x) = \dfrac{1}{x^2 + a^2}, \quad a > 0, \quad \hat{u}(\xi) = \dfrac{\pi e^{-a\xi}}{a}$,

    (d) $u(x) = e^{-ax^2}, \quad a > 0, \quad \hat{u}(\xi) = \dfrac{\sqrt{\pi}}{\sqrt{a}} e^{-\xi^2/4a}$,

    (e) $u(x) = \dfrac{\sin ax}{x}, \quad a > 0, \quad \hat{u}(\xi) = \begin{cases} \pi, & |\xi| < a, \\ \frac{\pi}{2}, & |\xi| = a, \\ 0, & |\xi| > a. \end{cases}$

11. Derive the following basic properties of Fourier transform (here $\hat{u} = \mathcal{F}\{u\}$):

    (a) $\mathcal{F}\{x^n u(x)\} = i^n \hat{u}^{(n)}(\xi)$,

    (b) $\mathcal{F}\{e^{iax} u(x)\} = \hat{u}(\xi - a)$,

    (c) $\mathcal{F}\{u(x - a)\} = e^{-ia\xi} \hat{u}(\xi)$,

    (d) $\mathcal{F}\{u(ax)\} = \dfrac{1}{|a|} \hat{u}(\dfrac{\xi}{a}), \quad a \neq 0$.

12. Using the Fourier transform method, solve the following Cauchy problems.

    (a)
    $$\begin{cases} u_{tt} = u_{xx}, & x \in \mathbb{R}, \; t > 0, \\ u(x, 0) = \dfrac{1}{1 + x^2}, \; u_t(x, 0) = 0. \end{cases}$$

    $[u(x,t) = \frac{1}{2} \int_{-\infty}^{\infty} e^{-|\xi|} \cos \xi t \, e^{i\xi x} d\xi]$

    (b)
    $$\begin{cases} u_t = \dfrac{1}{100} u_{xx}, & x \in \mathbb{R}, \; t > 0, \\ u(x, 0) = \varphi(x), \end{cases}$$

    where
    $$\varphi(x) = \begin{cases} 100, & -1 < x < 1, \\ 0, & \text{elsewhere.} \end{cases}$$

    (c)
    $$\begin{cases} u_{tt} = c^2 u_{xx}, & x \in \mathbb{R}, \; t > 0, \\ u(x, 0) = \sqrt{\dfrac{2}{\pi}} \dfrac{\sin x}{x}, \; u_t(x, 0) = 0. \end{cases}$$

    $[u(x,t) = \frac{50}{\sqrt{\pi t}} \int_{-\infty}^{\infty} \frac{1}{1+\xi^2} e^{-(x-\xi)^2/4t} d\xi]$

(d)
$$\begin{cases} u_t = u_{xx}, & x \in \mathbb{R},\ t > 0, \\ u(x,0) = \varphi(x), \end{cases}$$

where
$$\varphi(x) = \begin{cases} 1 - |x|/2, & -2 < x < 2, \\ 0, & \text{elsewhere.} \end{cases}$$

(e)
$$\begin{cases} u_t = e^{-t} u_{xx}, & x \in \mathbb{R},\ t > 0, \\ u(x,0) = 100. \end{cases}$$

13. Using the Fourier transform, solve the linearized Korteweg-deVries equation
$$u_t = u_{xxx}, \quad x \in \mathbb{R},\ t > 0,$$
subject to the initial condition
$$u(x,0) = e^{-x^2/2}.$$

14. Using the Fourier transform, solve the Cauchy problem
$$\begin{cases} u_{tt} = a^2 u_{txx} - b u_{xxxx}, & x \in \mathbb{R},\ t > 0, \\ u(x,0) = \varphi(x), \\ u_t(x,0) = \psi(x). \end{cases}$$

15. Using the Fourier transform, solve the heat equation with a convection term
$$u_t = k u_{xx} + \mu u_x, \quad x \in \mathbb{R},\ t > 0,$$
with an initial condition $u(x,0) = \varphi(x)$, assuming that $u(x,t)$ is bounded and $k > 0$.

$$[u(x,t) = \frac{1}{\sqrt{4\pi kt}} \int_{-\infty}^{\infty} \varphi(y) e^{-(\mu t + x - y)^2/(4kt)} dy]$$

16. Use the Fourier transform in the x variable to find the harmonic function in the half-plane $y > 0$ that satisfies the Neumann condition $\frac{\partial u}{\partial y} = h(x)$ on the boundary $y = 0$.

17. Use the Fourier transform to solve the Laplace equation $u_{xx} + u_{yy} = 0$ in the infinite strip $\{x \in \mathbb{R},\ 0 < y < 1\}$, together with the conditions $u(x,0) = 0$ and $u(x,1) = f(x)$.

$$[u(x,y) = \int_0^\infty \int_{-\infty}^\infty \frac{1}{\pi \sinh k} f(\xi) \sinh ky \cos(kx - k\xi)\, d\xi\, dk]$$

Exercises 4–6, 12 are taken from Asmar [3], 7, 8, 13, 14 from Keane [12], and 9, 15–17 from Strauss [19].

10 General Principles

In this chapter we summarize the main qualitative properties of the PDEs we dealt with in the previous chapters.

10.1 Principle of Causality (Wave Equation)

As we already know from d'Alembert's formula and Chapter 4, the values of the initial displacement φ and the initial velocity ψ at a point x_0 influence the solution of the wave equation only in the *domain of influence*, which is a sector determined by the characteristic lines $x \pm ct = x_0$ (see Figure 10.1).

Conversely, a solution at a point (x, t) is influenced only by the values from the *domain of dependence*, which is formed by the *characteristic triangle* with vertices $(x - ct, 0)$, $(x + ct, 0)$ and (x, t) (see Figure 10.1).

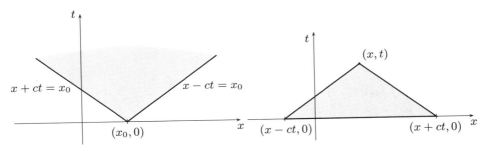

Figure 10.1 *Domain of influence of the point $(x_0, 0)$ and domain of dependence of the point (x, t).*

However, these properties follow directly from the wave equation itself and the knowledge of the formula for the solution is not needed. To prove it, we proceed in the following way.

We start with the wave equation $u_{tt} - c^2 u_{xx} = 0$ and multiply it by u_t. The resulting identity can be written as

(10.1)
$$\begin{aligned} 0 &= u_{tt} u_t - c^2 u_{xx} u_t \\ &= (\frac{1}{2} u_t^2 + \frac{1}{2} c^2 u_x^2)_t - c^2 (u_t u_x)_x \\ &= (\partial_x, \partial_t) \cdot (-c^2 u_t u_x, \frac{1}{2} u_t^2 + \frac{1}{2} c^2 u_x^2). \end{aligned}$$

Notice that the last scalar product is a two-dimensional divergence of the vector $\boldsymbol{f} := (-c^2 u_t u_x, \frac{1}{2} u_t^2 + \frac{1}{2} c^2 u_x^2)$. Now, we integrate (10.1) over a trapezoid F, which is part of the characteristic triangle (see Figure 10.2). If we use Green's Theorem, which can be written as

$$\iint_F \text{div}\, \boldsymbol{f}\, dx\, dy = \int_{\partial F} \boldsymbol{f} \cdot \boldsymbol{n}\, ds$$

(cf. its other version used in Section 4.3), we obtain

(10.2) $$\int_{\partial F} [(-c^2 u_t u_x) n_1 + (\tfrac{1}{2} u_t^2 + \tfrac{1}{2} c^2 u_x^2) n_2]\, ds = 0.$$

Here $\boldsymbol{n} = (n_1, n_2)$ is an outer normal vector to ∂F. The boundary ∂F consists of "top" T, "bottom" B, and "sides" $K = K_1 \cup K_2$. Thus, the integral in (10.2) splits into four parts

$$\int_{\partial F} = \int_T + \int_B + \int_{K_1} + \int_{K_2} = 0.$$

Now, we consider each part separately. On the top T, the normal vector is $\boldsymbol{n} = (0,1)$ and thus

$$\int_T \boldsymbol{f} \cdot \boldsymbol{n}\, ds = \int_T \tfrac{1}{2}(u_t^2 + c^2 u_x^2)\, ds.$$

On the bottom B we have $\boldsymbol{n} = (0,-1)$ and thus

$$\int_B \boldsymbol{f} \cdot \boldsymbol{n}\, ds = \int_B -\tfrac{1}{2}(u_t^2 + c^2 u_x^2)\, ds = \int_B -\tfrac{1}{2}(\psi^2 + c^2 \varphi_x^2)\, ds.$$

On the side K_1 there is $\boldsymbol{n} = (1, c)$ and

$$\int_{K_1} \boldsymbol{f} \cdot \boldsymbol{n}\, ds = \int_{K_1} [c\tfrac{1}{2}(u_t^2 + c^2 u_x^2) - c^2 u_t u_x]\, ds = \int_{K_1} \tfrac{c}{2}(u_t - c u_x)^2\, ds \geq 0.$$

Similarly, on K_2 we have $\boldsymbol{n} = (-1, c)$ and

$$\int_{K_1} \boldsymbol{f} \cdot \boldsymbol{n}\, ds = \int_{K_1} [c\tfrac{1}{2}(u_t^2 + c^2 u_x^2) + c^2 u_t u_x]\, ds = \int_{K_1} \tfrac{c}{2}(u_t + c u_x)^2\, ds \geq 0.$$

Putting all these partial results together, we can conclude that

$$\int_T \tfrac{1}{2}(u_t^2 + c^2 u_x^2)\, ds - \int_B \tfrac{1}{2}(\psi^2 + c^2 \varphi_x^2)\, ds \leq 0$$

or, equivalently,

(10.3) $$\int_T (u_t^2 + c^2 u_x^2)\, ds \leq \int_B (\psi^2 + c^2 \varphi_x^2)\, ds.$$

If we now assume that both the functions φ and ψ are zero on B, inequality (10.3) implies $u_t^2 + c^2 u_x^2 = 0$ on T, and thus $u_t \equiv u_x \equiv 0$ on T. Moreover, since this result holds true for a trapezoid of any height, we obtain that u_t and u_x are zero (and thus u is constant) in the whole characteristic triangle. And since we have assumed $u \equiv 0$ on B, we can conclude that $u \equiv 0$ in the whole triangle.

This result also implies uniqueness: if we take two solutions u_1 and u_2 with the same initial conditions on B, i.e., in the interval $(x_0 - ct_0, x_0 + ct_0)$, then $u_1 \equiv u_2$ in the whole characteristic triangle.

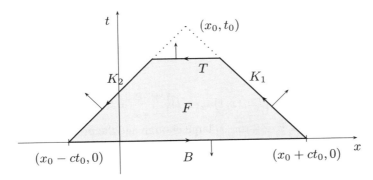

Figure 10.2 *Trapezoid of characteristic triangle.*

The principle of causality can be obtained also by other (and probably simpler) methods. However, the advantage of this approach is its applicability in any dimension (see Section 13.5). Notice that the analogue of Green's Theorem in two-dimensional case is the so called Divergence Theorem in higher dimensions.

10.2 Energy Conservation Law (Wave Equation)

Let us start with the infinitely long string described by the equation

(10.4) $$\rho u_{tt}(x,t) = T u_{xx}(x,t), \quad x \in \mathbb{R},\ t > 0,$$

where ρ, T are constants (we assume the displacements to be small enough). Moreover, let us consider for simplicity $\varphi \equiv \psi \equiv 0$ for $|x| > R$, $R > 0$ large enough. We suppose that the Cauchy problem has a classical solution.

The kinetic energy of the string takes the form

$$E_k(t) = \frac{1}{2} \int_{-\infty}^{\infty} \rho u_t^2(x,t)\, dx,$$

(cf. the well-known formula $E_k = \text{``}\frac{1}{2}mv^2\text{''}$ for the kinetic energy of the mass point).

The continuity assumption on u_{tt} implies

$$\frac{dE_k(t)}{dt} = \rho \int_{-\infty}^{\infty} u_t(x,t) u_{tt}(x,t)\, dx,$$

and after substituting for u_{tt} from equation (10.4) we obtain

(10.5) $$\frac{dE_k(t)}{dt} = T \int_{-\infty}^{\infty} u_t(x,t) u_{xx}(x,t)\, dx$$

$$= \Big[T u_t(x,t) u_x(x,t)\Big]_{x=-\infty}^{x=+\infty} - T \int_{-\infty}^{\infty} u_{tx}(x,t) u_x(x,t)\, dx.$$

Since

$$\Big[T u_t(x,t) u_x(x,t)\Big]_{x=-\infty}^{x=+\infty} = 0$$

(cf. the assumptions $\varphi \equiv \psi \equiv 0$ for $|x|$ large enough and the principle of causality) and

$$u_{tx}(x,t) u_x(x,t) = \left(\frac{1}{2} u_x^2(x,t)\right)_t,$$

we obtain, after substituting into (10.5),

(10.6) $$\frac{dE_k(t)}{dt} = -\frac{d}{dt} \int_{-\infty}^{\infty} \frac{1}{2} T u_x^2(x,t)\, dx.$$

The potential energy of the string can be expressed as

(10.7) $$E_p(t) = \frac{1}{2} T \int_{-\infty}^{\infty} u_x^2(x,t)\, dx.$$

Remark 10.1 This formula can be derived, for instance, in the following way. The potential energy represents the product of the force and the extension caused by this force (cf. the well-known formula $E_p =$ "mgh" for the potential energy of the mass point). In our case, the acting force is represented by the tension T. The extension h in the case of the string of length l is $h = s - l$, where s is the arc length, thus

$$h(t) = \int_0^l \sqrt{1 + u_x^2(x,t)}\, dx - l.$$

If we replace the square root on the right-hand side by the first two terms of its Taylor expansion, we obtain

$$h(t) \approx \int_0^l [1 + \frac{1}{2} u_x^2(x,t)]\, dx - l = \frac{1}{2} \int_0^l u_x^2(x,t)\, dx.$$

Section 10.2 Energy Conservation Law (Wave Equation)

The potential energy is then $E_p(t) = \frac{1}{2}T \int_0^l u_x^2(x,t)\,dx$. For the string of infinite length, we obtain expression (10.7).

Relations (10.6) and (10.7) imply

$$\frac{dE_k(t)}{dt} = -\frac{dE_p(t)}{dt}.$$

Since the total string energy is

$$E(t) = E_k(t) + E_p(t),$$

we obtain

$$\frac{dE(t)}{dt} = 0,$$

which is—in the language of mathematics—the *energy conservation law*. In other words, the total string energy

$$\boxed{E(t) = \frac{1}{2}\int_{-\infty}^{\infty} (\rho u_t^2(x,t) + T u_x^2(x,t))\,dx \equiv E}$$

is constant with respect to t!

Example 10.2 Let us determine the total energy of the infinitely long string, if the initial velocity at time $t = 0$ is zero and the initial displacement is given by the function

$$\varphi(x) = \begin{cases} b - \frac{b}{a}|x| & \text{for } |x| \le a, \\ 0 & \text{for } |x| > a. \end{cases}$$

Since the total string energy does not depend on time, we have

$$E = E(t) = E(0).$$

So, we need not find the solution at arbitrary time t, since the initial condition is sufficient for determination of the total energy. Zero initial velocity implies zero kinetic energy at time $t = 0$, thus

$$E_k(0) = 0.$$

The potential energy is expressed by

$$E_p(t) = \frac{1}{2}\int_{-\infty}^{\infty} T u_x^2(x,t)\,dx,$$

and hence

$$E_p(0) = \frac{1}{2}\int_{-\infty}^{\infty} T\varphi_x^2(x)\,dx = \frac{1}{2}\int_{-a}^{a} T\varphi_x^2(x)\,dx = \frac{1}{2}\int_{-a}^{a} T\frac{b^2}{a^2}\,dx = \frac{b^2}{a}T.$$

The total energy is then the sum of the potential and kinetic energies:

$$E = E(0) = \frac{b^2}{a}T.$$

□

10.3 Ill-Posed Problem (Diffusion Equation for Negative t)

First, let us consider the following initial value problem for a "special variant of the diffusion equation":

(10.8) $\qquad u_t = -u_{xx}, \quad u(x,0) = 1, \quad x \in \mathbb{R}, \ t > 0.$

Here, the diffusion coefficient k is equal to -1! Obviously, $u(x,t) \equiv 1$ is the solution. On the other hand, we can easily verify that the function

$$u_n(x,t) = 1 + \frac{1}{n}\sin nx \ e^{n^2 t}$$

solves the initial value problem

(10.9) $\qquad u_t = -u_{xx}, \quad u(x,0) = 1 + \frac{1}{n}\sin nx, \quad n = 1, 2, \ldots.$

The initial conditions in problems (10.8) and (10.9) differ only in the term $\frac{1}{n}\sin nx$, which converges to zero uniformly for $n \to \infty$. However, the difference of the corresponding solutions is

$$\frac{1}{n}\sin nx \ e^{n^2 t},$$

which for any fixed x (except integer multiples of π) goes to infinity for $n \to \infty$. It means that the *stability of the constant solution* $u(x,t) \equiv 1$ *fails* for the diffusion equation with a negative coefficient and such a problem is ill-posed.

Now, let us consider the initial value problem for the diffusion equation ($k > 0$) with negative time,

(10.10) $\qquad u_t = ku_{xx}, \quad u(x,0) = \varphi(x), \quad x \in \mathbb{R}, \ t < 0.$

Using the substitution

$$w(x,t) = u(x,-t),$$
$$w_t = -u_t, \quad w_{xx} = u_{xx},$$

we obtain the problem

(10.11) $\qquad w_t = -kw_{xx}, \quad w(x,0) = \varphi(x), \quad x \in \mathbb{R}, \ t > 0.$

However, we have shown above that such a problem (for $k = 1$) is ill-posed. So, problem (10.10) is ill-posed as well.

This phenomenon has also its physical explanation: diffusion, heat flow, the so called Brownian motion, etc. are *irreversible processes* and return in time leads to *chaos*. On the other hand, the wave motion is a reversible process and the wave equation for $t < 0$ is well-posed.

10.4 Maximum Principle (Heat Equation)

Let us now leave the problems on the whole real line and consider the heat flow in a finite bar of length l, whose ends are held at time t at temperatures $g(t)$ and $h(t)$, respectively, while the initial distribution of temperature at time $t = 0$ is given by a continuous function $\varphi(x)$. We have thus an initial boundary value problem

(10.12)
$$\begin{array}{l} u_t = ku_{xx}, \quad x \in (0,l), \; t > 0, \\ u(x,0) = \varphi(x), \\ u(0,t) = g(t), \; u(l,t) = h(t). \end{array}$$

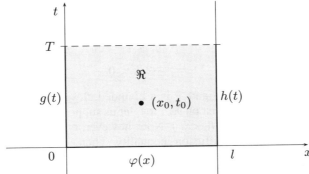

Figure 10.3 *Space-time cylinder $\Re$.*

We introduce a notion of a space-time cylinder $\Re$, by which we understand, in our case, a rectangle in the xt-plane whose one vertex is placed at the origin $(0,0)$, and two sides lie on coordinate axes (see Figure 10.3). The length of side on the x-axis is l, the length of side on the t-axis is T. By the bottom of the cylinder $\Re$ we understand the horizontal side lying on the x-axis, by the cylinder jacket we understand both lateral sides. The upper horizontal line is then called the top of the cylinder $\Re$. The reason why we use these terms is that the maximum principle can be derived in the same way also for the diffusion equation in more spatial variables (where the idea of the cylinder is more realistic).

We will prove the following assertion.

Theorem 10.3 *Let $u = u(x,t)$ be a classical solution of problem (10.12). Then u achieves its extremal values (minimal as well as maximal) on the bottom or jacket of the space-time cylinder $\Re$.*

Actually, a stronger assertion holds true. It says that the values of the function u inside the cylinder and on the top are *strictly* less (or greater) than the maximum (or minimum, respectively) on the rest of the boundary of the cylinder $\Re$ (unless the function $u(x,t)$ is constant)—see, e.g., Protter, Weinberger [16].

Proof (of Theorem 10.3) The proof will be done for the maximum value. In the case of the minimum value, we would proceed analogously (using the fact that $\min u = -\max(-u)$).

The idea of the proof uses the fact that the first partial derivatives of the function must be zero and the second derivatives must be non-positive at the *inner* maximum point.

Let us denote by M the maximal value of the function $u(x,t)$ on the sides $t=0$, $x=0$, $x=l$, and let us put $v(x,t) = u(x,t) + \epsilon x^2$, where ϵ is a positive constant. Our goal is to show that $v(x,t) \leq M + \epsilon l^2$ on the whole cylinder $\Re$.

The definition of the function v implies that the inequality $v(x,t) \leq M + \epsilon l^2$ is satisfied on the boundary lines $t=0$, $x=0$ and $x=l$. Further, the so called *diffusion inequality* $v_t(x,t) - kv_{xx}(x,t) < 0$ holds for all $(x,t) \in \Re$. Indeed,

$$\begin{aligned} v_t(x,t) - kv_{xx}(x,t) &= u_t(x,t) - k(u(x,t) + \epsilon x^2)_{xx} \\ &= u_t(x,t) - ku_{xx}(x,t) - 2\epsilon k \\ &= -2\epsilon k < 0. \end{aligned}$$

Since v is a continuous function and $\Re$ is a bounded closed set, the maximum point (x_0, t_0) of the function v must exist on $\Re$. First, let us suppose that this point lies *inside* the cylinder $\Re$ ($0 < x_0 < l$, $0 < t_0 < T$). Now, however, $v_t = 0$, $v_{xx} \leq 0$ must hold at (x_0, t_0), which contradicts the diffusion inequality. Then, let (x_0, t_0) lie on the upper side of the space-time cylinder $\Re$, that is, $t_0 = T$, $0 < x_0 < l$. Thus $v_x(x_0, t_0) = 0$, $v_{xx}(x_0, t_0) \leq 0$ and $v_t(x_0, t_0) \geq 0$, which again contradicts the diffusion inequality. Consequently, the maximum point of the function v must be achieved on the rest of the boundary: on the lines $t=0$, $x=0$, $x=l$. Here, however, we have the inequality $v(x,t) \leq M + \epsilon l^2$. These facts imply that the relation $v(x,t) \leq M + \epsilon l^2$ holds for all $(x,t) \in \Re$. If we substitute for v, we obtain $u(x,t) \leq M + \epsilon(l^2 - x^2)$. Since $\epsilon > 0$ has been chosen completely arbitrarily, it follows that

$$u(x,t) \leq M$$

for any $(x,t) \in \Re$, which we wanted to prove.

■

Remark 10.4 We have met a variant of Maximum Principle also in the case of the diffusion equation on the whole real line. The maximal as well as minimal values of the solution were achieved at time $t=0$ and, with the growing time, the solution values were "spread" and tended to some value between those extremes (see Example 5.4).

Corollary 10.5 (Uniqueness.) *The initial boundary value problem (10.12) has at most one classical solution.*

Proof Let $u_1(x,t)$ and $u_2(x,t)$ be two classical solutions of problem (10.12). Let us put $w = u_1 - u_2$. Then the function w satisfies

$$w_t - kw_{xx} = 0, \quad w(x,0) = 0, \quad w(0,t) = 0, \quad w(l,t) = 0.$$

According to the maximum principle, the function w achieves its maximum on the sides $t = 0$, $x = 0$, $x = l$, where it is, however, equal to zero. Thus $w(x,t) \leq 0$. The same argument for the minimum value yields $w(x,t) \geq 0$. Hence, we obtain $w(x,t) \equiv 0$, and thus $u_1(x,t) \equiv u_2(x,t)$.

∎

Corollary 10.6 (Uniform stability.) *Let u_1, u_2 be two classical solutions of the initial boundary value problem (10.12) corresponding to two initial conditions φ_1, φ_2. Then*

$$\max_{x \in [0,l]} |u_1(x,t) - u_2(x,t)| \leq \max_{x \in [0,l]} |\varphi_1(x) - \varphi_2(x)|$$

for each $t > 0$. In particular, the classical solution of (10.12) is stable with respect to small perturbations of the initial condition.

This statement says that "small" change of the initial condition results in a "small" change of the solution at arbitrary time.

Proof Let the solution $u_1(x,t)$ corresponds to the initial condition $\varphi_1(x)$, the solution $u_2(x,t)$ corresponds to the initial condition $\varphi_2(x)$; boundary conditions are the same in both cases. Again, let us denote by $w = u_1 - u_2$ the difference of the two solutions. The function w solves the problem

$$w_t - kw_{xx} = 0, \quad w(x,0) = \varphi_1(x) - \varphi_2(x), \quad w(0,t) = 0, \quad w(l,t) = 0.$$

The maximum principle then implies

$$w(x,t) = u_1(x,t) - u_2(x,t) \leq \max_{x \in [0,l]} (\varphi_1 - \varphi_2) \leq \max_{x \in [0,l]} |\varphi_1 - \varphi_2|.$$

Similarly, according to the "minimum principle" (i.e., the maximum principle applied to $-u$),

$$w(x,t) = u_1(x,t) - u_2(x,t) \geq \min_{x \in [0,l]} (\varphi_1 - \varphi_2) \geq - \max_{x \in [0,l]} |\varphi_1 - \varphi_2|.$$

Consequently,

$$\max_{x \in [0,l]} |u_1(x,t) - u_2(x,t)| \leq \max_{x \in [0,l]} |\varphi_1(x) - \varphi_2(x)|$$

for each $t > 0$.

∎

10.5 Energy Method (Diffusion Equation)

We will show another way how to prove uniqueness of the classical solution of problem (10.12) and its stability (now, however, with respect to a more general norm). The technique of the proof is called the *energy method*.

Let us consider again two solutions $u_1(x,t)$, $u_2(x,t)$ of problem (10.12) and their difference $w(x,t)$. The function w satisfies the homogeneous diffusion equation with homogeneous boundary conditions $w(0,t) = w(l,t) = 0$. If we multiply the diffusion equation for w by the function w itself, we obtain

$$0 = (w_t - kw_{xx})w = (\frac{1}{2}w^2)_t + (-kw_x w)_x + kw_x^2.$$

Integrating over the interval $0 < x < l$, we get

$$0 = \int_0^l (\frac{1}{2}w^2(x,t))_t \, \mathrm{d}x - kw_x(x,t)w(x,t)\Big|_{x=0}^{x=l} + k \int_0^l w_x^2(x,t) \, \mathrm{d}x.$$

The second term on the right-hand side vanishes due to the zero boundary conditions. Changing the order of time differentiation and integration, we obtain

$$\frac{\mathrm{d}}{\mathrm{d}t} \int_0^l \frac{1}{2} w^2(x,t) \, \mathrm{d}x = -k \int_0^l w_x^2(x,t) \, \mathrm{d}x \leq 0.$$

But this means that the integral depending on the parameter t, $\int_0^l w^2(x,t) \, \mathrm{d}x$, is—as a function of time t—decreasing, and thus

(10.13) $$\int_0^l w^2(x,t) \, \mathrm{d}x \leq \int_0^l w^2(x,0) \, \mathrm{d}x$$

for $t \geq 0$. In the case that both solutions $u_1(x,t)$, $u_2(x,t)$ correspond to the same initial condition, we obtain $w(x,0) = 0$, and hence $\int_0^l w^2(x,t) \, \mathrm{d}x = 0$ for all $t > 0$. This means that $w \equiv 0$ and hence $u_1 \equiv u_2$ for all $t \geq 0$. In other words, we obtain again uniqueness of the solution of the initial boundary value problem (10.12).

If the solutions $u_1(x,t)$, $u_2(x,t)$ correspond to different initial conditions $\varphi_1(x)$, $\varphi_2(x)$, respectively, then $w(x,0) = \varphi_1(x) - \varphi_2(x)$ and relation (10.13) becomes

$$\int_0^l (u_1(x,t) - u_2(x,t))^2 \, \mathrm{d}x \leq \int_0^l (\varphi_1(x) - \varphi_2(x))^2 \, \mathrm{d}x,$$

which expresses *stability of the solution with respect to the initial condition in the L^2-norm*.

10.6 Maximum Principle (Laplace Equation)

One of the fundamental properties of all harmonic functions is the *Maximum Principle*.

Section 10.6 Maximum Principle (Laplace Equation)

Theorem 10.7 (Maximum Principle.) *Let Ω be a bounded domain (i.e., an open and connected set) in $\mathbb{R}^2$. Let $u(x,y)$ be a harmonic function on Ω (that is, $\Delta u = 0$ in Ω) which is continuous on $\overline{\Omega} = \Omega \cup \partial\Omega$. Then the maximal and minimal values of the function u are achieved on the boundary $\partial\Omega$.*

Like for the diffusion equation, we can formulate a stronger assertion. We will state it later and use the Poisson formula for its proof, see Section 10.7. For now, however, we put up with the weaker version formulated in Theorem 10.7 and give its elementary proof, which is somewhat similar to that of Theorem 10.3.

Proof For simplicity, let us denote $\boldsymbol{x} = (x, y)$ and $|\boldsymbol{x}| = (x^2 + y^2)^{1/2}$. We introduce $v(\boldsymbol{x}) = u(\boldsymbol{x}) + \epsilon|\boldsymbol{x}|^2$, where ϵ is an arbitrarily small positive constant. We have

$$\Delta v = \Delta u + \epsilon \Delta(x^2 + y^2) = 0 + 4\epsilon > 0$$

in the whole domain Ω. If the function v achieved its maximum inside Ω, the inequality $\Delta v = v_{xx} + v_{yy} \leq 0$ would have to hold at such a point, but this contradicts the previous inequality. Thus the maximum v must be achieved at a point on the boundary—let us denote this point $\boldsymbol{x}_0 \in \partial\Omega$. Then, for all $\boldsymbol{x} \in \Omega$, we obtain

$$u(\boldsymbol{x}) \leq v(\boldsymbol{x}) \leq v(\boldsymbol{x}_0) = u(\boldsymbol{x}_0) + \epsilon|\boldsymbol{x}_0|^2 \leq \max_{\boldsymbol{y} \in \partial\Omega} u(\boldsymbol{y}) + \epsilon|\boldsymbol{x}_0|^2.$$

Since ϵ has been chosen arbitrary, we have

$$u(\boldsymbol{x}) \leq \max_{\boldsymbol{y} \in \partial\Omega} u(\boldsymbol{y}) \qquad \forall \boldsymbol{x} \in \overline{\Omega} = \Omega \cup \partial\Omega.$$

For the case of minimum, we proceed analogously. ■

Theorem 10.8 (Uniqueness of the solution of the Dirichlet problem.) *The solution of the Dirichlet problem for the Poisson equation in the domain Ω is uniquely determined.*

Proof Let us assume that

$$\begin{cases} \Delta u = f & \text{in } \Omega, \\ u = h & \text{on } \partial\Omega, \end{cases} \qquad \begin{cases} \Delta v = f & \text{in } \Omega, \\ v = h & \text{on } \partial\Omega. \end{cases}$$

If we put $w = u - v$, we obtain $\Delta w = 0$ in Ω and $w = 0$ on $\partial\Omega$. Theorem 10.7 implies

$$0 = \min_{\boldsymbol{y} \in \overline{\Omega}} w(\boldsymbol{y}) \leq w(\boldsymbol{x}) \leq \max_{\boldsymbol{y} \in \overline{\Omega}} w(\boldsymbol{y}) = 0 \qquad \text{for all } \boldsymbol{x} \in \Omega.$$

However, this means that $w \equiv 0$ and thus $u \equiv v$. ■

10.7 Consequences of Poisson Formula (Laplace Equation)

An important result following from the Poisson formula is the so called *Mean Value Property Theorem*.

Theorem 10.9 (Mean Value Property.) *Let u be a harmonic function on a disc D, continuous in its closure $\overline{D}$. Then the value of the function u at the center of D is equal to the mean value of u on the circle ∂D.*

Proof We shift the coordinate system so that the origin $\mathbf{0}$ is placed at the center of the disc (this can be done because the Laplace operator is invariant with respect to translations). We substitute $\boldsymbol{x} = \mathbf{0}$ into the Poisson formula (8.17) obtaining

$$u(\mathbf{0}) = \frac{a^2}{2\pi a} \int_{|\boldsymbol{x}'|=a} \frac{u(\boldsymbol{x}')}{a^2}\,ds = \frac{1}{2\pi a}\int_{\partial D} u(\boldsymbol{x}')\,ds.$$

This is exactly the integral mean value of the function u over the circle $|\boldsymbol{x}'| = a$. ∎

Another important consequence of the Poisson formula is the strong version of the Maximum Principle.

Theorem 10.10 (Strong Maximum Principle.) *Let u be a harmonic function in the domain $\Omega \in \mathbb{R}^2$, continuous on $\overline{\Omega}$. Then either u is constant in the entire closure $\overline{\Omega}$, or u achieves its maximal (minimal) value only on the boundary $\partial \Omega$ (i.e., never inside Ω).*

Idea of proof Let us denote by $\boldsymbol{x}_M$ the point where u achieves its maximal value M on the closure of the domain Ω (its existence follows from the continuity of u on $\overline{\Omega}$ and from the Weierstrass Theorem). We shall show that $\boldsymbol{x}_M$ cannot lie inside Ω unless u is constant.

Let $\boldsymbol{x}_M \in \Omega$. Let us consider a circle centered at the point $\boldsymbol{x}_M$ which is the boundary of a circular neighborhood of $\boldsymbol{x}_M$ entirely contained in Ω. According to the Mean Value Theorem, $u(\boldsymbol{x}_M)$ equals the mean value of u over the circle. Since $M := u(\boldsymbol{x}_M)$ is the maximal value, all values of u on the circle must be equal to M as well. Moreover, the same holds for an arbitrary circle with the center $\boldsymbol{x}_M$ and smaller radius. Thus $u(\boldsymbol{x}) = M$ for all $\boldsymbol{x}$ from the original circular neighborhood. Now, we can imagine that we cover the whole domain Ω by circular neighborhoods (see Figure 10.4). Since Ω is connected, we obtain $u(\boldsymbol{x}) \equiv M$ on the whole domain Ω, and thus u is constant. ∎

The last consequence of the Poisson formula which we state here is the following differentiability assertion.

Section 10.7 Consequences of Poisson Formula (Laplace Equation)

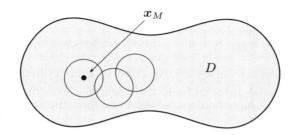

Figure 10.4 *Covering of D by circular neighborhoods.*

Theorem 10.11 (Differentiability.) *Let u be a harmonic function on an open set $\Omega \subset \mathbb{R}^2$. Then $u(\boldsymbol{x}) = u(x, y)$ has on Ω continuous partial derivatives of all orders.*

This property of harmonic functions is—in a certain sense—similar to the property that we have met when studying the diffusion equation (see Chapter 5).

Idea of proof First, let us consider an open disc D with the center at the origin. In the Poisson formula (8.17), the integrand is a function having partial derivatives of arbitrary orders for all $\boldsymbol{x} \in D$. Notice that $\boldsymbol{x}' \in \partial D$ and thus $\boldsymbol{x} \neq \boldsymbol{x}'$. Since we can change the order of integration and differentiation, the function u has partial derivatives of all orders in D as well.

Now, let us denote by D a circular neighborhood of the point $\boldsymbol{x}_0 \in \Omega$ which is wholly contained in Ω. Using the substitution $\boldsymbol{y} := \boldsymbol{x} - \boldsymbol{x}_0$, we translate the center of D in the origin. It then follows from above that u is differentiable in D. Since $\boldsymbol{x}_0 \in \Omega$ is arbitrary, u is differentiable at all points of Ω. ∎

Remark 10.12 (Laplace Equation in Finite Differences.) Let us have a look at the Laplace equation from the "numerical" point of view. Let us consider a point (x, y) and its neighbors $(x \pm h, y)$, $(x, y \pm h)$, where $h > 0$ is small enough. The Taylor expansion yields

$$u(x - h, y) = u(x, y) - h u_x(x, y) + \frac{1}{2} h^2 u_{xx}(x, y) + O(h^3),$$

$$u(x + h, y) = u(x, y) + h u_x(x, y) + \frac{1}{2} h^2 u_{xx}(x, y) + O(h^3)$$

and, after summation, we can express the second derivative in the form

$$u_{xx}(x, y) = \frac{1}{h^2} [u(x - h, y) - 2u(x, y) + u(x + h, y)] + O(h^2).$$

Similarly,

$$u_{yy}(x,y) = \frac{1}{h^2}[u(x,y-h) - 2u(x,y) + u(x,y+h)] + O(h^2).$$

Notice that the second derivatives are expressed by central differences used, for instance, in the grid method. If we substitute in the Laplace equation $\Delta u = u_{xx} + u_{yy} = 0$ and neglect the terms of higher orders, we obtain an approximate value of the function u at the point (x,y) as

$$u(x,y) \approx \frac{1}{4}[u(x-h,y) + u(x+h,y) + u(x,y-h) + u(x,y+h)].$$

However, it means that the value $u(x,y)$ is approximately the arithmetic average of the surrounding values. This arithmetic average cannot be neither greater nor less than all the surrounding values. Thus, even here, we meet a certain numerical analogue of the Maximum Principle and the Mean Value Theorem.

10.8 Comparison of Wave, Diffusion and Laplace Equations

The above mentioned principles and properties of solutions of initial and boundary value problems yield the following fundamental comparison of all three types of linear equations of the second order (cf. Strauss [19]):

Property	Wave	Diffusion	Laplace
Speed of propagation	Finite ($\leq c$)	Infinite	Zero
Singularities for $t > 0$	Propagate along characteristics at speed c	Disappear immediately (solutions are regular)	Solutions are regular
Well-posedness	Yes for $t > 0$ Yes for $t < 0$	Yes for $t > 0$ No for $t < 0$	Yes
Maximum principle	No	Yes	Yes
Behavior for $t \to \infty$ (for equations on $\mathbb{R}, \mathbb{R}^2$)	Energy does not decrease (is constant)	Energy decreases (if φ is integrable)	Steady state

10.9 Exercises

1. Show that the wave equation has the following invariance properties:

Section 10.9 Exercises

(a) Any shifted solution $u(x - y, t)$, where y is fixed, is also a solution.

(b) Any derivative of the solution (e.g., u_x) is also a solution.

(c) Dilated solution $u(ax, at)$, where $a > 0$, is also a solution.

2. For a solution $u(x, t)$ of the wave equation (10.4) with $\rho = T = c = 1$, the energy density is defined as $e = \frac{1}{2}(u_t^2 + u_x^2)$ and the momentum density as $p = u_t u_x$.

(a) Show that $e_t = p_x$ and $p_t = e_x$.

(b) Show that both $e(x, t)$ and $p(x, t)$ also satisfy the wave equation.

3. Let $u(x, t)$ solve the wave equation $u_{tt} = u_{xx}$. Prove that the identity

$$u(x + h, t + k) + u(x - h, t - k) = u(x + k, t + h) + u(x - k, t - h)$$

holds for all x, t, k and h. Draw the characteristic parallelogram with vertices formed by the arguments in the previous relation.

4. Consider a *damped* infinite string described by the equation $u_{tt} - c^2 u_{xx} + r u_t = 0$, and show that its total energy decreases.

5. Consider two Cauchy problems for the wave equation with different initial data:

$$\begin{cases} u_{tt}^i = c^2 u_{xx}^i, & x \in \mathbb{R},\ 0 < t < T, \\ u^i(x, 0) = \varphi^i(x), & u_t^i(x, 0) = \psi^i(x) \end{cases}$$

for $i = 1, 2$, where φ^1, φ^2, ψ^1, ψ^2 are given functions. If

$$|\varphi^1(x) - \varphi^2(x)| \leq \delta_1, \quad |\psi^1(x) - \psi^2(x)| \leq \delta_2$$

for all $x \in \mathbb{R}$, show that $|u^1(x, t) - u^2(x, t)| \leq \delta_1 + \delta_2 T$ for all $x \in \mathbb{R}$, $0 < t < T$. What does it mean with regard to stability?

6. Using the energy conservation law for the wave equation, prove that the initial value problem

$$u_{tt} = c^2 u_{xx}, \quad x \in \mathbb{R},\ t > 0,$$
$$u(x, 0) = \varphi(x),\ u_t(x, 0) = \psi(x)$$

has a unique solution.

7. Consider the diffusion equation on the real line. Using the Maximum Principle, show that an odd (even) initial condition leads to an odd (even) solution.

8. Consider a solution of the diffusion equation $u_t = u_{xx}$, $0 \leq x \leq l$, $0 \leq t < \infty$.

(a) Let $M(T)$ be the maximum of the function $u(x, t)$ on the rectangle $\{0 \leq x \leq l, 0 \leq t \leq T\}$. Is $M(T)$ decreasing or as a increasing function of T?

(b) Let $m(T)$ be the minimum of the function $u(x, t)$ on the rectangle $\{0 \leq x \leq l, 0 \leq t \leq T\}$. Is $m(T)$ decreasing or increasing as a function of T?

9. Consider the diffusion equation $u_t = u_{xx}$ on the interval $(0, 1)$ with boundary conditions $u(0, t) = u(1, t) = 0$ and the initial condition $u(x, 0) = 1 - x^2$. Notice that the initial condition does not satisfy the boundary condition on the left end, however, the solution satisfies it at arbitrary time $t > 0$.

   (a) Show that $u(x, t) > 0$ at all inner points $0 < x < 1$, $0 < t < \infty$.

   (b) Let, for all $t > 0$, $\mu(t)$ represent the maximum of the function $u(x, t)$ on $0 \leq x \leq 1$. Show that $\mu(t)$ is a non-increasing function of t.

   [Hint: Suppose the maximum to be achieved at a point $X(t)$, i.e., $\mu(t) = u(X(t), t)$. Differentiate $\mu(t)$ (under the assumption that $X(t)$ is a differentiable function).]

   (c) Sketch the solution on several time levels.

10. Consider the diffusion equation $u_t = u_{xx}$ on the interval $(0, 1)$ with boundary conditions $u(0, t) = u(1, t) = 0$ and the initial condition $u(x, 0) = 4x(1 - x)$.

    (a) Show that $0 < u(x, t) < 1$ for all $t > 0$ and $0 < x < 1$.

    (b) Show that $u(x, t) = u(1 - x, t)$ for all $t \geq 0$ and $0 \leq x \leq 1$.

    (c) Using the energy method (see Section 10.5), show that $\int_0^1 u^2(x, t)\,dx$ is a strictly decreasing function of t.

11. The aim of this exercise is to show that the maximum principle does not hold true for the equation $u_t = xu_{xx}$ which has a variable coefficient.

    (a) Verify that the function $u(x, t) = -2xt - x^2$ is a solution. Find its maximum on the rectangle $\{-2 \leq x \leq 2,\ 0 \leq t \leq 1\}$.

    (b) Where exactly does our proof of the maximum principle fail in the case of this equation?

12. Consider a heat problem with an internal heat source
$$\begin{cases} u_t = u_{xx} + 2(t + 1)x + x(1 - x), & 0 < x < 1,\ t > 0, \\ u(0, t) = 0,\quad u(1, t) = 0, \\ u(x, 0) = x(1 - x). \end{cases}$$

    Show that the maximum principle does not hold true:

    (a) Verify that $u(x, t) = (t + 1)x(1 - x)$ is a solution.

    (b) What are the maximum value M and the minimum value m of the initial and boundary data?

    (c) Show that, for some $t > 0$, the temperature distribution exceeds M at certain points of the bar.

13. Prove the comparison principle for the diffusion equation: If u and v are two solutions and $u \leq v$ for $t = 0$, for $x = 0$ and $x = l$, then $u \leq v$ for $0 \leq t < \infty$, $0 \leq x \leq l$.

14. (a) More generally, if $u_t - ku_{xx} = f$, $v_t - kv_{xx} = g$, $f \leq g$ and $u \leq v$ for $x = 0$, $x = l$ and $t = 0$, then $u \leq v$ for $0 \leq t < \infty$, $0 \leq x \leq l$. Prove it.

    (b) Let $v_t - v_{xx} \geq \sin x$ for $0 \leq x \leq \pi$, $0 < t < \infty$. Further, let $v(0,t) \geq 0$, $v(\pi, t) \geq 0$ and $v(x, 0) \geq \sin x$. Exploit part (a) for proving that $v(x, t) \geq (1 - e^{-t}) \sin x$.

15. Consider the diffusion equation on $(0, l)$ with the Robin boundary conditions $u_x(0, t) - a_0 u(0, t) = 0$ and $u_x(l, t) + a_l u(l, t) = 0$. If $a_0 > 0$ and $a_l > 0$, use the energy method to show that the endpoints contribute to the decrease of $\int_0^l u^2(x, t)dx$. (Part of the energy is lost at the boundary, so the boundary conditions are called *radiating* or *dissipative*.)

16. Here we show the direct relation between the wave and diffusion equations. Let $u(x, t)$ solve the wave equation on the whole real line, and let its second derivatives be bounded. Put

$$v(x,t) = \frac{c}{\sqrt{4\pi kt}} \int_{-\infty}^{\infty} e^{-s^2c^2/(4kt)} u(x, s)\, ds.$$

    (a) Show that $v(x, t)$ solves the diffusion equation.

    (b) Show that $\lim_{t \to 0} v(x, t) = u(x, 0)$.

    [Hint: (a) Write the formula in the form $v(x, t) = \int_{-\infty}^{\infty} H(s, t) u(x, s)\, ds$, where $H(x, t)$ solves the diffusion equation with constant k/c^2 for $t > 0$. Then differentiate $v(x, t)$. (b) Use the fact that $H(s, t)$ is a fundamental solution of the diffusion equation with the spatial variable s.]

17. Show that there is no maximum principle for the wave equation.

18. Let u be a harmonic function in the disc $D = \{r < 2\}$ and let $u = 3 \sin 2\theta + 1$ for $r = 2$. Without finding the concrete form of the solution, answer the following questions:

    (a) What is the maximal value of u on $\overline{D}$?

    (b) What is the value of u at the origin?

    [(a) 4, (b) 1]

19. Prove uniqueness of the Dirichlet problem $\Delta u = f$ in D, $u = g$ on ∂D by the energy method.

20. Let Ω be a bounded open set and consider the Neumann problem

$$\Delta u = f \text{ in } \Omega, \quad \frac{\partial u}{\partial n} = g \text{ on } \partial \Omega.$$

Show that any two solutions differ by a constant.

21. Let Ω be a bounded open set. Show that the Neumann problem
$$\Delta u + \alpha u = f \text{ in } \Omega, \quad \frac{\partial u}{\partial n} = g \text{ on } \partial\Omega$$
has at most one solution if $\alpha < 0$ in Ω.

22. Prove that the function $u(x,y) = \frac{1-x^2-y^2}{x^2+(y-1)^2}$ is harmonic in $\mathbb{R}^2 \setminus \{0,1\}$. Find the maximum M and the minimum m of $u(x,y)$ in the disc $D_\rho = \{x^2+y^2 < \rho^2\}$, $\rho < 1$, and show that $Mm = 1$. Plot the graph of $u(x,y)$, where $(x,y) \in D_{0.9}$, using polar coordinates.

Exercises 1–4, 8–11, 13–19 are taken from Strauss [19], 5 from Logan [14], 12 from Asmar [3], 20, 21 from Keane [12], and 22 from Stavroulakis and Tersian [18].

11 Laplace and Poisson equations in Higher Dimensions

In this chapter we treat the Laplace operator and harmonic functions in $\mathbb{R}^3$. Unlike the two-dimensional case, we cannot rely on methods based on direct computation of the solution, since the situation is much more complicated. That is why we try to obtain as much information as possible about the solution and its properties from the equation itself. In particular, we focus on those aspects that differ from the two-dimensional case. Features that do not depend on the dimension are stated without details.

The fundamental property of harmonic functions that—together with the proof technique—does not depend on the dimension is the *Weak Maximum Principle*. We recommend the reader to study Theorem 10.7 and its proof in order to realize its applicability in higher dimensions. Also the consequence concerning *uniqueness of the solution of the Dirichlet problem* (see Theorem 10.8) can be stated without any changes.

11.1 Invariance of the Laplace Operator and its Transformation into Spherical Coordinates

The Laplace operator is invariant with respect to translations and rotations in three as well as in two dimensions. Let us recall that, using the matrix notation, rotation in $\mathbb{R}^3$ is given by the transformation formula

$$x' = Bx,$$

where $x = (x, y, z)$ and $B = (b_{ij})$, $i, j = 1, 2, 3$, is an *orthogonal matrix* (that is, $BB^t = B^tB = I$). Using the chain rule we derive

$$u_x = b_{11}u_{x'} + b_{21}u_{y'} + b_{31}u_{z'},$$
$$u_y = b_{12}u_{x'} + b_{22}u_{y'} + b_{32}u_{z'},$$
$$u_z = b_{13}u_{x'} + b_{23}u_{y'} + b_{33}u_{z'},$$
$$u_{xx} = b_{11}^2 u_{x'x'} + b_{11}b_{21}u_{x'y'} + b_{11}b_{31}u_{x'z'}$$
$$+ b_{21}b_{11}u_{y'x'} + b_{21}^2 u_{y'y'} + b_{21}b_{31}u_{y'z'}$$
$$+ b_{31}b_{11}u_{z'x'} + b_{31}b_{21}u_{z'y'} + b_{31}^2 u_{z'z'}$$

and similar formulas for u_{yy} and u_{zz}. Summing up, we obtain

$$u_{xx} + u_{yy} + u_{zz} = \underbrace{(b_{11}^2 + b_{12}^2 + b_{13}^2)}_{=1} u_{x'x'}$$
$$+ 2\underbrace{(b_{11}b_{21} + b_{12}b_{22} + b_{13}b_{23})}_{=0} u_{x'y'} + 2\underbrace{(b_{11}b_{31} + b_{12}b_{32} + b_{13}b_{33})}_{=0} u_{x'z'}$$
$$+ \cdots = u_{x'x'} + u_{y'y'} + u_{z'z'}$$

due to the orthogonality of the matrix B. The reader is invited to carry out all the above calculations in detail.

The rotational invariancy of the Laplace operator implies that, in radially symmetric cases, the transformation into *spherical coordinates* (r, θ, ϕ) could bring a significant simplification (see Figure 11.1).

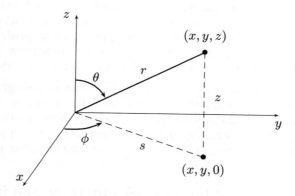

Figure 11.1 *Spherical coordinates.*

We will use the notation

$$r = \sqrt{x^2 + y^2 + z^2} = \sqrt{s^2 + z^2},$$
$$s = \sqrt{x^2 + y^2},$$
$$x = s \cos \phi, \quad z = r \cos \theta,$$
$$y = s \sin \phi, \quad s = r \sin \theta.$$

For the transformation, we use the knowledge of the transform relations into the polar coordinates:

$$u_{zz} + u_{ss} = u_{rr} + \frac{1}{r} u_r + \frac{1}{r^2} u_{\theta\theta},$$

$$u_{xx} + u_{yy} = u_{ss} + \frac{1}{s} u_s + \frac{1}{s^2} u_{\phi\phi}.$$

Summing up and canceling u_{ss}, we obtain

$$\Delta u = u_{xx} + u_{yy} + u_{zz} = u_{rr} + \frac{1}{r} u_r + \frac{1}{r^2} u_{\theta\theta} + \frac{1}{s} u_s + \frac{1}{s^2} u_{\phi\phi}.$$

In the last term we insert $s^2 = r^2 \sin^2 \theta$. Moreover, u_s can be written as

$$u_s = \frac{\partial u}{\partial s} = u_r \frac{\partial r}{\partial s} + u_\theta \frac{\partial \theta}{\partial s} + u_\phi \frac{\partial \phi}{\partial s}.$$

Section 11.2 Green's First Identity

Evidently, $\frac{\partial \phi}{\partial s} = 0$. If we use, e.g., the inverse Jacobi matrix of the transformation into polar coordinates (cf. Section 6.2 with $r = r$, $\theta = \theta$ and $s = y$), we obtain

$$\frac{\partial r}{\partial s} = \sin\theta = \frac{s}{r}, \qquad \frac{\partial \theta}{\partial s} = \frac{\cos\theta}{r},$$

and thus

$$u_s = u_r \frac{s}{r} + u_\theta \frac{\cos\theta}{r}.$$

Hence, we easily derive

$$\Delta u = u_{rr} + \frac{2}{r} u_r + \frac{\cos\theta}{r^2 \sin\theta} u_\theta + \frac{1}{r^2} u_{\theta\theta} + \frac{1}{r^2 \sin^2\theta} u_{\phi\phi}.$$

Written in symbols, we obtain the following analogue of the "two-dimensional formula" (6.1):

(11.1)
$$\boxed{\begin{aligned} \Delta &= \frac{\partial^2}{\partial x^2} + \frac{\partial^2}{\partial y^2} + \frac{\partial^2}{\partial z^2} \\ &= \frac{\partial^2}{\partial r^2} + \frac{2}{r}\frac{\partial}{\partial r} + \frac{\cos\theta}{r^2 \sin\theta}\frac{\partial}{\partial \theta} + \frac{1}{r^2}\frac{\partial^2}{\partial \theta^2} + \frac{1}{r^2 \sin^2\theta}\frac{\partial^2}{\partial \phi^2}. \end{aligned}}$$

In the radially symmetric situation, that is, when u does not depend on the angles ϕ and θ, the Laplace equation in spherical coordinates reduces to the ODE

$$\Delta u = u_{rr} + \frac{2}{r} u_r = 0.$$

Thus, $(r^2 u_r)_r = 0$. Hence we obtain $u_r = c_1 / r^2$ and $u = -c_1 r^{-1} + c_2$. For $c_1 = -1$, $c_2 = 0$, the harmonic function

$$\boxed{u(r) = \frac{1}{r} = (x^2 + y^2 + z^2)^{-\frac{1}{2}}}$$

is the analogue of the two-dimensional harmonic function $\ln(x^2 + y^2)^{1/2}$, which we met in Chapter 6. In *electrostatics*, for instance, $u(x) = r^{-1}$ represents the *electrostatic potential* which corresponds to the unit charge placed at the origin.

11.2 Green's First Identity

In the sequel, we focus on the three-dimensional case; however, all statements remain valid even in the two-dimensional case or, generally, in any dimension.

The main tool we will use is the *Divergence Theorem*, which can be expressed by the relation

$$\iiint_\Omega \operatorname{div} \mathbf{F} \, d\mathbf{x} = \iint_{\partial\Omega} \mathbf{F} \cdot \mathbf{n} \, dS,$$

(under certain assumptions concerning the vector function $\boldsymbol{F}$, the bounded domain (open and connected set) $\Omega \subset \mathbb{R}^3$ and its boundary $\partial\Omega$). From the physical point of view, this theorem says that the contribution of all sources and sinks in the given domain (the left-hand side) must be equal to the sum of all inflows and outflows across the boundary (the right-hand side). The symbol $\boldsymbol{n}$ denotes the unit vector of the outer normal to $\partial\Omega$ (see Figure 11.2).

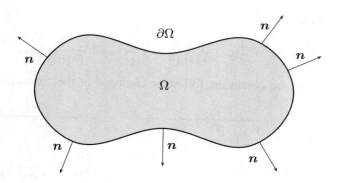

Figure 11.2 *Domain Ω with the unit vector of the outer normal to the boundary.*

Green's first identity. According to the product rule, we write

$$\begin{aligned}(vu_x)_x &= v_x u_x + v u_{xx}, \\ (vu_y)_y &= v_y u_y + v u_{yy}, \\ (vu_z)_z &= v_z u_z + v u_{zz}.\end{aligned}$$

Summing up all these three equations, we obtain

$$\nabla \cdot (v \nabla u) = \nabla v \cdot \nabla u + v \Delta u.$$

If we integrate this relation and use the Divergence Theorem for the left-hand side, we obtain *Green's first identity*

(11.2) $$\boxed{\iint_{\partial\Omega} v \frac{\partial u}{\partial n}\, \mathrm{d}S = \iiint_\Omega \nabla v \cdot \nabla u\, \mathrm{d}\boldsymbol{x} + \iiint_\Omega v \Delta u\, \mathrm{d}\boldsymbol{x},}$$

where $\partial u/\partial n = \boldsymbol{n} \cdot \nabla u$ is the derivative with respect to the outer normal to the boundary of the domain Ω. The identity (11.2) can be interpreted as a three-dimensional version of integration by parts and has a number of consequences.

11.3 Properties of Harmonic Functions

The first consequence of Green's first identity is the three-dimensional version of the *Mean Value Property* (we have met its two-dimensional variant in Theorem 10.9).

Theorem 11.1 (Mean Value Property.) *The average value of any harmonic function in a domain $\Omega \subset \mathbb{R}^3$ over any sphere which lies in Ω is equal to its value at the center of the sphere.*

Proof Following, e.g., Strauss [19] or Zauderer [22], let us consider a ball $B(\mathbf{0}, a) = \{|x| < a\} \subset \Omega$ with radius a centered at the origin (recall the invariancy of the Laplace operator with respect to translations). Then $\partial B(\mathbf{0}, a) = \{|x| = a\}$. Further, let $\Delta u = 0$ in Ω, $\partial B(\mathbf{0}, a) \subset \Omega$. For the sphere, the outer normal $\mathbf{n}$ has the direction of the radius vector, thus

$$\frac{\partial u}{\partial n} = \mathbf{n} \cdot \nabla u = \frac{\mathbf{x}}{r} \cdot \nabla u = \frac{x}{r} u_x + \frac{y}{r} u_y + \frac{z}{r} u_z$$

$$= \frac{\partial x}{\partial r} u_x + \frac{\partial y}{\partial r} u_y + \frac{\partial z}{\partial r} u_z = \frac{\partial u}{\partial r},$$

where $r = (x^2 + y^2 + z^2)^{1/2} = |x|$ is the spherical coordinate (the distance of the point (x, y, z) from the center $\mathbf{0}$ of the sphere). If we use Green's first identity for the ball $B(\mathbf{0}, a)$ with the choice $v \equiv 1$, we obtain

$$\iint_{\partial B(\mathbf{0},a)} \frac{\partial u}{\partial r} \, dS = \iiint_{B(\mathbf{0},a)} \Delta u \, dx = 0.$$

We rewrite the integral on the left-hand side into spherical coordinates (r, θ, ϕ), that is,

$$\int_0^{2\pi} \int_0^{\pi} u_r(a, \theta, \phi) a^2 \sin \theta \, d\theta \, d\phi = 0$$

(on the sphere $\partial B(\mathbf{0}, a)$, we have $r = a$), and divide the equality by the constant $4\pi a^2$, which is the measure of $\partial B(\mathbf{0}, a)$ (that is, the surface of the ball $B(\mathbf{0}, a)$). This result holds true for all $a > 0$, thus we can replace a with the variable r. Moreover, if we change the order of integration and differentiation (which is possible under certain assumptions on u), we obtain

(11.3) $$\frac{\partial}{\partial r} \left(\frac{1}{4\pi} \int_0^{2\pi} \int_0^{\pi} u(r, \theta, \phi) \sin \theta \, d\theta \, d\phi \right) = 0.$$

However, it means that the expression

$$\frac{1}{4\pi} \int_0^{2\pi} \int_0^{\pi} u(r, \theta, \phi) \sin \theta \, d\theta \, d\phi$$

which represents the average value of u on the sphere $\{|x| = r\}$, is *independent of the radius r*. In particular, for $r \to 0$ we have

(11.4) $$\frac{1}{4\pi} \int_0^{2\pi} \int_0^{\pi} u(\mathbf{0}) \sin\theta \, d\theta \, d\phi = u(\mathbf{0}).$$

Relations (11.3) and (11.4) imply the following result for any $a > 0$:

(11.5) $$\boxed{u(\mathbf{0}) = \frac{1}{\operatorname{meas} \partial B(\mathbf{0}, a)} \iint_{\partial B(\mathbf{0},a)} u \, dS.}$$

Thus, the proof of the Mean Value Property in three dimensions is completed. (Observe that the idea of the proof can be applied generally in any dimension.) ∎

Like in two dimensions (see Theorem 10.10), a direct consequence of the Mean Value Property is the *strong version of the Maximum Principle*.

Theorem 11.2 (Strong Maximum Principle.) *Let Ω be an arbitrary domain in $\mathbb{R}^3$. A non-constant harmonic function in Ω, continuous in $\overline{\Omega}$, cannot achieve its maximum (minimum) inside Ω, but only on the boundary $\partial\Omega$.*

The proof follows the same lines as that of Theorem 10.10, which is why we do not repeat it here.

Another important theorem following from Green's first identity that has also physical motivation, is the *Dirichlet Principle*.

Theorem 11.3 (Dirichlet Principle.) *Let $u(x)$ be the harmonic function on a domain Ω satisfying the Dirichlet boundary condition*

(11.6) $$u(\mathbf{x}) = h(\mathbf{x}) \quad \text{on } \partial\Omega.$$

Let $w(\mathbf{x})$ be an arbitrary continuously differentiable function on $\overline{\Omega}$ satisfying (11.6). Then

$$E(w) \geq E(u),$$

where E denotes the energy defined by the formula

(11.7) $$\boxed{E(w) = \frac{1}{2} \iiint_{\Omega} |\nabla w|^2 \, d\mathbf{x}.}$$

Section 11.3 Properties of Harmonic Functions

In other words, the Dirichlet Principle says that, among all functions satisfying the boundary condition (11.6), the harmonic function corresponds to the state with the lowest energy. Expression (11.7) represents just the *potential energy*—there is no motion and thus the kinetic energy is zero. One of the fundamental physical principles says that any system tends to keep the state with the lowest energy. Harmonic functions thus describe the most frequent "ground (quiescent) states".

Proof (of Theorem 11.3) Let us denote $w = u + v$ and convert the definition formula for the energy (11.7) in the following way:

$$E(w) = \frac{1}{2} \iiint_\Omega |\nabla(u+v)|^2 \, d\boldsymbol{x}$$

$$= E(u) + \iiint_\Omega \nabla u \cdot \nabla v \, d\boldsymbol{x} + E(v).$$

To the middle term we apply Green's first identity and use the fact that $v = 0$ on $\partial\Omega$ (both u and w satisfy the same Dirichlet boundary condition) and $\Delta u = 0$ in Ω. Consequently, $\iiint_\Omega \nabla u \cdot \nabla v \, d\boldsymbol{x} = 0$ and

$$E(w) = E(u) + E(v).$$

Since, evidently, $E(v) \geq 0$, we obtain $E(w) \geq E(u)$ and the Dirichlet Principle is proved. ∎

Uniqueness of Solution of Dirichlet Problem. A direct consequence of the Maximum Principle is the uniqueness of the solution of the Dirichlet problem for the Poisson equation. We refer the reader to Theorem 10.8 and its proof that can be applied in any dimension. Here we present another technique based on the so called *energy method*, which consists in the application of Green's first identity.

Let us consider the Dirichlet problem $\Delta u = f$ in the domain Ω, $u = h$ on the boundary $\partial\Omega$, and a couple of solutions u_1, u_2. We denote their difference by $u = u_1 - u_2$. The function u is harmonic in Ω, vanishing on the boundary $\partial\Omega$. Now, we use Green's first identity (11.2) for $v = u$. Since u is a harmonic function ($\Delta u = 0$ in Ω), we obtain

$$\iint_{\partial\Omega} u \frac{\partial u}{\partial n} \, dS = \iiint_\Omega |\nabla u|^2 \, d\boldsymbol{x}.$$

Since $u = 0$ on the boundary $\partial\Omega$, the left-hand side is equal to zero. This yields

$$\iiint_\Omega |\nabla u|^2 \, d\boldsymbol{x} = 0,$$

which implies $|\nabla u| = 0$ in Ω. This means that the function u is constant in the domain Ω. But since it vanishes on the boundary $\partial\Omega$, we obtain $u(\boldsymbol{x}) \equiv 0$ in Ω, and thus $u_1(\boldsymbol{x}) \equiv u_2(\boldsymbol{x})$ in Ω.

In a similar way we could prove that the solution of the *Neumann problem* is determined *uniquely up to a constant* (see Exercises 11.7).

Necessary Condition of Solvability of Neumann problem. If we use a special choice $v \equiv 1$, Green's first identity reads

$$(11.8) \qquad \iint_{\partial\Omega} \frac{\partial u}{\partial n}\,dS = \iiint_{\Omega} \Delta u\,d\boldsymbol{x}.$$

Let us consider the Neumann problem on the domain Ω

$$(11.9) \qquad \begin{cases} \Delta u = f(x) & \text{in } \Omega, \\ \frac{\partial u}{\partial n} = h(x) & \text{on } \partial\Omega \end{cases}$$

and let us substitute for $\frac{\partial u}{\partial n}$ and Δu into relation (11.8). We obtain a necessary condition for the solvability of (11.9) in the form

$$(11.10) \qquad \iint_{\partial\Omega} h\,dS = \iiint_{\Omega} f\,d\boldsymbol{x}.$$

It means that the solution of the Neumann problem can exist only if the input data (functions f and h) are not completely arbitrary but satisfy condition (11.10). In fact, it can be proved that condition (11.10) is also sufficient for problem (11.9) to have a solution. Thus, from the point of solvability, the Neumann problem is not well-posed. Concerning uniqueness, we find this problem ill-posed as well (if there exists a solution to (11.9) then adding an arbitrary constant to this solution, we obtain again a solution of (11.9)). Nevertheless, the Neumann boundary value problem makes reasonable sense and occurs very often in applications.

11.4 Green's Second Identity and Representation Formula

Green's second identity. If we apply Green's first identity to a couple of functions u and v, then to the couple v and u, and subtract the two equations, we obtain the relation

$$(11.11) \qquad \boxed{\iiint_{\Omega} (u\Delta v - v\Delta u)\,d\boldsymbol{x} = \iint_{\partial\Omega} \left(u\frac{\partial v}{\partial n} - v\frac{\partial u}{\partial n}\right) dS,}$$

which is known as *Green's second identity*.

An important consequence of Green's second identity is the so called *representation formula*. It says that the value of a harmonic function at any point of a domain Ω can be expressed using only its values on the boundary $\partial\Omega$.

Section 11.4 Green's Second Identity and Representation Formula

Theorem 11.4 (Representation Formula.) *Let $x_0 \in \Omega \subset \mathbb{R}^3$. The value of any harmonic function on a domain Ω, continuous on $\overline{\Omega}$, can be expressed by*

(11.12)
$$u(x_0) = \frac{1}{4\pi} \iint_{\partial\Omega} \left[-u(x) \frac{\partial}{\partial n} \left(\frac{1}{|x-x_0|} \right) + \frac{1}{|x-x_0|} \frac{\partial u}{\partial n}(x) \right] dS.$$

Observe that relation (11.12) contains the fundamental radially symmetric harmonic function $r^{-1} = |x-x_0|^{-1}$ that we have already met in the previous sections of this chapter (here it is shifted by the vector x_0).

Proof Relation (11.12) is a special case of Green's second identity for the choice

$$v(x) = \frac{-1}{4\pi|x-x_0|}.$$

This function is, however, *unbounded* at x_0, thus we cannot use Green's second identity on the whole domain Ω. Let us denote by Ω_ε the domain $\Omega \setminus B(x_0, \varepsilon)$, where $B(x_0, \varepsilon) \subset \Omega$ is the ball centered at the point x_0 with radius ε. This domain is now admissible for the application of Green's second identity.

For simplicity, let us shift the point x_0 to the origin (we recall again the invariancy of the Laplace operator with respect to translations). Hence, $v(x) = -1/(4\pi r)$, where $r = |x| = (x^2 + y^2 + z^2)^{1/2}$. If we use the fact that $\Delta u = \Delta v = 0$ in Ω_ε, Green's second identity implies

$$-\frac{1}{4\pi} \iint_{\partial\Omega_\varepsilon} \left[u \frac{\partial}{\partial n} \left(\frac{1}{r} \right) - \frac{\partial u}{\partial n} \frac{1}{r} \right] dS = 0.$$

The boundary $\partial\Omega_\varepsilon$ consists of two parts: the boundary of the original domain $\partial\Omega$ and the sphere $\partial B(x_0, \varepsilon)$. Moreover, on this sphere we have $\partial/\partial n = -\partial/\partial r$. The surface integral above is thus decomposed into two parts and the equality can be written as

(11.13)
$$-\frac{1}{4\pi} \iint_{\partial\Omega} \left(u \frac{\partial}{\partial n} \left(\frac{1}{r} \right) - \frac{\partial u}{\partial n} \frac{1}{r} \right) dS = -\frac{1}{4\pi} \iint_{r=\varepsilon} \left(u \frac{\partial}{\partial r} \left(\frac{1}{r} \right) - \frac{\partial u}{\partial r} \frac{1}{r} \right) dS.$$

This equality must hold true for any (small) $\varepsilon > 0$. Concerning the sphere $|x| = r = \varepsilon$, we have

$$\frac{\partial}{\partial r} \left(\frac{1}{r} \right) = -\frac{1}{r^2} = -\frac{1}{\varepsilon^2}.$$

The right-hand side of relation (11.13) can be thus rewritten in the form

$$\frac{1}{4\pi\varepsilon^2} \iint_{r=\varepsilon} u\, dS + \frac{1}{4\pi\varepsilon} \iint_{r=\varepsilon} \frac{\partial u}{\partial r} dS = \overline{u} + \varepsilon \overline{\frac{\partial u}{\partial r}},$$

where $\bar{u}$ denotes the integral average value of the function $u(x)$ on the sphere $|x| = r = \varepsilon$, and $\overline{\partial u/\partial r}$ represents the average value of $\partial u/\partial n$ on this sphere. If we now pass to the limit for $\varepsilon \to 0$, we obtain

$$\bar{u} + \varepsilon \overline{\frac{\partial u}{\partial r}} \longrightarrow u(\mathbf{0}) + 0 \times \frac{\partial u}{\partial r}(\mathbf{0}) = u(\mathbf{0})$$

(note that the function u is continuous and $\partial u/\partial r$ is bounded). Hence, from relation (11.13), we easily get formula (11.12). (The reader is asked to give the reasons.) ∎

Remark 11.5 In the same way we could obtain the corresponding formula in any dimension. The concrete form of this formula in N dimensions depends on the corresponding fundamental radially symmetric harmonic function which, for $N \geq 3$, has the form r^{-N+2}, and for $N = 2$ is equal to $\ln r$ (see Section 6.3). In particular, in *two dimensions*, the representation formula reads

$$\boxed{u(x_0) = \frac{1}{2\pi} \int_{\partial\Omega} \left(u(x) \frac{\partial}{\partial n} (\ln|x - x_0|) - \frac{\partial u}{\partial n}(x) \ln|x - x_0| \right) ds,}$$

where $\Delta u = 0$ in the plane domain Ω and $x_0 \in \Omega$. We integrate here along the curve $\partial\Omega$, and ds denotes an element of the arc length of this curve.

11.5 Boundary Value Problems and Green's Function

The main disadvantage of the representation formula (11.12) is that it contains the boundary values of both the functions u and $\frac{\partial u}{\partial n}$. But solving the standard boundary value problems, we are usually given either the Dirichlet boundary condition or the Neumann boundary condition, not both at the same time! The representation formula is based on two properties of the function $v(x) = -1/(4\pi|x - x_0|)$: it is a harmonic function except the point x_0, and the singularity at this point has a "proper" form. Our goal is to modify this function in such a way that we could eliminate one term in formula (11.12). The modified function will be called *Green's function* corresponding to the domain Ω.

Definition 11.6 *Green's function* $G(x, x_0)$ corresponding to the Laplace operator, the homogeneous Dirichlet boundary condition, a domain Ω and a point $x_0 \in \Omega$ is a function defined by the following properties:

(i) $G(x, x_0)$ has continuous second partial derivatives with respect to x and $\Delta G = 0$ in Ω except the point $x = x_0$ (here, the Laplace operator is considered with respect to x, while x_0 is a parameter).

(ii) $G(x, x_0) = 0$ for $x \in \partial\Omega$.

Section 11.5 Boundary Value Problems and Green's Function

(iii) The function $G(\boldsymbol{x}, \boldsymbol{x}_0) + 1/(4\pi|\boldsymbol{x} - \boldsymbol{x}_0|)$ is finite at the point $\boldsymbol{x}_0$, has continuous partial derivatives of the second order in the whole domain Ω, and is harmonic.

It can be proved that Green's function exists and is determined uniquely (the uniqueness proof is based on the Maximum Principle, or on the theorem on the unique solvability of the Dirichlet problem; we let the details to the reader—see Exercises 11.7).

Theorem 11.7 *If $G(\boldsymbol{x}, \boldsymbol{x}_0)$ is Green's function, then the solution of the Dirichlet problem for the Laplace equation can be expressed by*

(11.14)
$$u(\boldsymbol{x}_0) = \iint_{\partial\Omega} u(\boldsymbol{x}) \frac{\partial G(\boldsymbol{x}, \boldsymbol{x}_0)}{\partial n} \, \mathrm{d}S.$$

Proof The representation formula implies

(11.15)
$$u(\boldsymbol{x}_0) = \iint_{\partial\Omega} \left(u \frac{\partial v}{\partial n} - \frac{\partial u}{\partial n} v \right) \mathrm{d}S,$$

where again $v(\boldsymbol{x}) = -(4\pi|\boldsymbol{x} - \boldsymbol{x}_0|)^{-1}$. Now, we define $H(\boldsymbol{x}) = G(\boldsymbol{x}, \boldsymbol{x}_0) - v(\boldsymbol{x})$. According to the property (iii) of Definition 11.6, the function $H(\boldsymbol{x})$ is harmonic on the whole domain Ω. We can thus apply Green's second identity to the couple $u(\boldsymbol{x})$, $H(\boldsymbol{x})$:

(11.16)
$$0 = \iint_{\partial\Omega} \left(u \frac{\partial H}{\partial n} - \frac{\partial u}{\partial n} H \right) \mathrm{d}S.$$

Summing (11.15) and (11.16), we obtain

$$u(\boldsymbol{x}_0) = \iint_{\partial\Omega} \left(u \frac{\partial G}{\partial n} - \frac{\partial u}{\partial n} G \right) \mathrm{d}S.$$

Moreover, according to (ii), Green's function satisfies $G = 0$ on the boundary $\partial\Omega$. This directly implies (11.14). ∎

Remark 11.8 Green's function is symmetric, that is,

$$G(\boldsymbol{x}, \boldsymbol{x}_0) = G(\boldsymbol{x}_0, \boldsymbol{x}) \quad \text{for } \boldsymbol{x} \neq \boldsymbol{x}_0.$$

In electrostatics, $G(\boldsymbol{x}, \boldsymbol{x}_0)$ represents the electric potential inside a closed conductive surface $S = \partial\Omega$ induced by a charge placed at the point $\boldsymbol{x}_0$. The symmetry of Green's function is then known as the *Reciprocity Principle*, according to which the source placed at $\boldsymbol{x}_0$ causes the same effect at the point $\boldsymbol{x}$ as the source at $\boldsymbol{x}$ causes at the point $\boldsymbol{x}_0$.

Green's function can be exploited also for solving the *Poisson equation*.

Theorem 11.9 *The Dirichlet boundary value problem for the Poisson equation*

(11.17)
$$\begin{cases} \Delta u = f & \text{in } \Omega, \\ u = h & \text{on } \partial\Omega \end{cases}$$

has a unique solution given by the formula

(11.18)
$$\boxed{u(\boldsymbol{x}_0) = \iint_{\partial\Omega} h(\boldsymbol{x}) \frac{\partial G(\boldsymbol{x}, \boldsymbol{x}_0)}{\partial n}\, \mathrm{d}S + \iiint_{\Omega} f(\boldsymbol{x}) G(\boldsymbol{x}, \boldsymbol{x}_0)\, \mathrm{d}\boldsymbol{x}}$$

for any $\boldsymbol{x}_0 \in \Omega$.

Uniqueness of the solution was discussed above in Section 11.3. It is straightforward to verify that $u = u(\boldsymbol{x})$ given by (11.18) is a solution of (11.17)—see Exercises 11.7.

The disadvantage of relations (11.14), (11.18) is the necessity to know the explicit expression of Green's function. This is possible only on domains with special geometry. Two such cases are considered in the forthcoming section (cf. Strauss [19], or Stavroulakis, Tersian [18]).

11.6 Dirichlet Problem on Half-Space and on Ball

The half-space and the ball in $\mathbb{R}^3$ belong among the domains for which Green's function and, consequently, the solution of the corresponding Dirichlet problem can be found explicitly. In both cases we use the so called *reflection method*.

Dirichlet Problem on Half-Space. Although the half-space is an unbounded domain, all assertions stated above—including the notion of Green's function—remain valid, provided we add the "boundary condition at infinity". By this condition, we understand the assumption that functions and their derivatives vanish for $|\boldsymbol{x}| \to \infty$.

We denote the coordinates of the point $\boldsymbol{x}$ by (x, y, z) as usual. The half-space $\Omega = \{\boldsymbol{x}, z > 0\}$ is the domain lying "above" the xy-plane. To each point $\boldsymbol{x} = (x, y, z) \in \Omega$ there corresponds the *reflected point* $\boldsymbol{x}^* = (x, y, -z)$ that evidently does not lie in Ω (see Figure 11.3).

We already know that the function $1/(4\pi|\boldsymbol{x} - \boldsymbol{x}_0|)$ satisfies the conditions (i) and (iii) imposed on Green's function. We try to modify it so as to ensure the validity of condition (ii).

We claim that Green's function for the half-space Ω has the form

(11.19)
$$\boxed{G(\boldsymbol{x}, \boldsymbol{x}_0) = -\frac{1}{4\pi|\boldsymbol{x} - \boldsymbol{x}_0|} + \frac{1}{4\pi|\boldsymbol{x} - \boldsymbol{x}_0^*|}.}$$

Section 11.6 Dirichlet Problem on Half-Space and on Ball

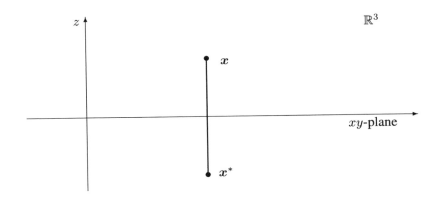

Figure 11.3 *Half-space and reflection method.*

Rewritten into coordinates, this becomes

$$G(\boldsymbol{x}, \boldsymbol{x}_0) = -\frac{1}{4\pi}[(x-x_0)^2 + (y-y_0)^2 + (z-z_0)^2]^{-1/2}$$
$$+ \frac{1}{4\pi}[(x-x_0)^2 + (y-y_0)^2 + (z+z_0)^2]^{-1/2}.$$

Observe that the two terms differ only by the element $(z \pm z_0)$. Let us verify that Green's function defined by formula (11.19) has the properties of Green's function stated in Definition 11.6.

(i) Obviously, G is finite and differentiable except the point $\boldsymbol{x}_0$. Also $\Delta G = 0$.

(ii) Let $\boldsymbol{x} \in \partial\Omega$, that is $z = 0$. Figure 11.4 illustrates that $|\boldsymbol{x} - \boldsymbol{x}_0| = |\boldsymbol{x} - \boldsymbol{x}_0^*|$. Hence, $G(\boldsymbol{x}, \boldsymbol{x}_0) = 0$ on $\partial\Omega$.

(iii) Since the point $\boldsymbol{x}_0^*$ lies outside the domain Ω, the function $-1/(4\pi|\boldsymbol{x} - \boldsymbol{x}_0^*|)$ has no singularities in Ω. The function G has thus a single singularity at the point $\boldsymbol{x}_0$ and this corresponds to the claims imposed on Green's function.

Thus, we have proved that formula (11.19) determines Green's function corresponding to the half-space Ω.

Now, we can use it for finding the solution of the Dirichlet problem

(11.20) $$\boxed{\begin{array}{l} \Delta u = 0 \quad \text{for } z > 0, \\ u(x, y, 0) = h(x, y). \end{array}}$$

We use formula (11.14). Notice that $\partial G/\partial n = -\partial G/\partial z|_{z=0}$, since the outer normal $\boldsymbol{n}$ has the "downward" direction (out of the domain). Further,

$$-\frac{\partial G}{\partial z} = \frac{1}{4\pi}\left(\frac{z+z_0}{|\boldsymbol{x} - \boldsymbol{x}_0^*|^3} - \frac{z-z_0}{|\boldsymbol{x} - \boldsymbol{x}_0|^3}\right) = \frac{1}{2\pi}\frac{z_0}{|\boldsymbol{x} - \boldsymbol{x}_0|^3}$$

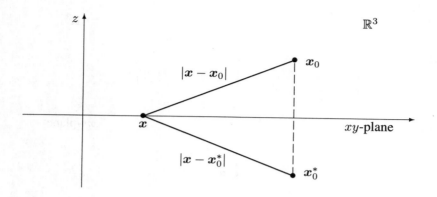

Figure 11.4 *Verification of the property (ii) of Green's function.*

for $z = 0$. Hence, by direct substitution, we obtain the solution of problem (11.20) in the form

$$u(x_0, y_0, z_0) = \frac{z_0}{2\pi} \int_{-\infty}^{\infty} \int_{-\infty}^{\infty} [(x-x_0)^2 + (y-y_0)^2 + z_0^2]^{-3/2} h(x,y) \, dx \, dy$$

or, in the vector notation,

(11.21) $$\boxed{u(\boldsymbol{x}_0) = \frac{z_0}{2\pi} \iint_{\partial\Omega} \frac{h(\boldsymbol{x})}{|\boldsymbol{x}-\boldsymbol{x}_0|^3} \, dS.}$$

Remark 11.10 We can proceed similarly in any dimension. In particular, let us have a look at the same problem in two dimensions, that is, let us consider the Laplace equation on the "upper half-plane":

(11.22) $$\boxed{\begin{array}{l} u_{xx} + u_{yy} = 0, \quad x \in \mathbb{R}, \ y > 0, \\ u(x,0) = h(x), \quad x \in \mathbb{R}. \end{array}}$$

The corresponding Green's function has the form

$$G(\boldsymbol{x}, \boldsymbol{x}_0) = \frac{1}{2\pi} \ln |\boldsymbol{x}-\boldsymbol{x}_0| - \frac{1}{2\pi} \ln |\boldsymbol{x}-\boldsymbol{x}_0^*|.$$

Here $\boldsymbol{x} = (x,y)$ and $|\boldsymbol{x}-\boldsymbol{x}_0| = \sqrt{(x-x_0)^2 + (y-y_0)^2}$. The solution of the Dirichlet

Section 11.6 Dirichlet Problem on Half-Space and on Ball

problem (11.22) is then given by

$$u(x_0, y_0) = \frac{y_0}{\pi} \int_{-\infty}^{\infty} \frac{h(x)}{|x - x_0|^2} \, dx = \frac{y_0}{\pi} \int_{-\infty}^{\infty} \frac{h(x)}{(x - x_0)^2 + y_0^2} \, dx.$$

Notice that problem (11.22) was solved with the same result also in Chapter 9; then, however, by the Fourier transform (see Remark 9.13)!

Dirichlet Problem on a Ball. Another domain where we can solve the Dirichlet problem using the explicitly found Green's function, is the ball $\Omega = \{|x| < a\}$ with radius a. Again we use the *reflection method*, this time, however, with respect to the sphere $\{|x| = a\}$ which forms the boundary $\partial \Omega$ (see Figure 11.5). The method is—in this case—called the *spherical inversion*.

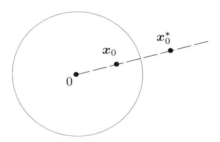

Figure 11.5 *Ball Ω and spherical inversion.*

Let us consider a fixed point $x_0 \in \Omega$. The *reflected point* x_0^* is determined by the following properties:

(i) x_0^* lies on the straight line passing through 0 and x_0,

(ii) its distance from the origin is given by the relation $|x_0| \, |x_0^*| = a^2$.

It means that
$$x_0^* = \frac{a^2 x_0}{|x_0|^2}.$$

Let x be an arbitrary point and let us denote $\rho(x) = |x - x_0|$ and $\rho^*(x) = |x - x_0^*|$. Then, for $x_0 \neq 0$, Green's function on the ball Ω is given by

(11.23) $$G(x, x_0) = -\frac{1}{4\pi\rho} + \frac{a}{|x_0|} \frac{1}{4\pi\rho^*}.$$

We prove this statement by verifying the properties (i), (ii) and (iii) of Definition 11.6. The case $x_0 = 0$ will be treated separately.

First of all, the single singularity of the function G is the point $x = x_0$, since x_0^* lies outside the ball Ω. Functions $1/\rho$ and $1/\rho^*$ are both harmonic in Ω except the point x_0. Conditions (i) and (iii) are thus fulfilled.

For the verification of condition (ii), we show that ρ^* is proportional to ρ for all x lying on the sphere $|x| = a$. The congruent triangles in Figure 11.6 imply

(11.24) $$\left|\frac{r_0}{a}x - \frac{a}{r_0}x_0\right| = |x - x_0|,$$

where $r_0 = |x_0|$. For the left-hand side of (11.24) we have

$$\frac{r_0}{a}\left|x - \frac{a^2}{r_0^2}x_0\right| = \frac{r_0}{a}\rho^*.$$

Hence, we obtain

$$\frac{r_0}{a}\rho^* = \rho \quad \text{for all } |x| = a.$$

However, this means that the function $G(x, x_0) = -\frac{1}{4\pi\rho} + \frac{a}{r_0}\frac{1}{4\pi\rho^*}$ is zero on the sphere $|x| = a$ and condition (ii) is satisfied.

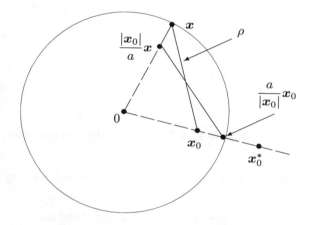

Figure 11.6 *Congruent triangles and the proportionality of ρ and ρ^*.*

Formula (11.23) can be rewritten to the form

(11.25) $$\boxed{G(x, x_0) = -\frac{1}{4\pi|x - x_0|} + \frac{1}{4\pi\left|\frac{r_0}{a}x - \frac{a}{r_0}x_0\right|}.}$$

In the case $x_0 = 0$, Green's function is

(11.26) $$\boxed{G(x, 0) = -\frac{1}{4\pi|x|} + \frac{1}{4\pi a}}$$

Section 11.6 Dirichlet Problem on Half-Space and on Ball

(verify—see Exercises 11.7).

Now, we use the knowledge of Green's function for finding the solution of the Dirichlet boundary value problem for the Laplace equation in the ball

(11.27) $$\boxed{\begin{aligned} \Delta u &= 0 \quad \text{for } |x| < a, \\ u &= h \quad \text{for } |x| = a. \end{aligned}}$$

We know from Theorem 11.1 (Mean Value Property) that $u(0)$ is the average value of the function $h(x)$ on the sphere $\partial\Omega$. Let us consider only the case $x_0 \neq 0$. Since we want to use the representation formula (11.14) (see Theorem 11.4), we have to determine $\partial G/\partial n$ on $|x| = a$. We start with the relation $\rho^2 = |x - x_0|^2$. Differentiating it with respect to x, we obtain $2\rho\nabla\rho = 2(x - x_0)$. Thus, $\nabla\rho = (x - x_0)/\rho$ and $\nabla(\rho^*) = (x - x_0^*)/\rho^*$. Now we determine the gradient of the function G from relation (11.23):

(11.28) $$\begin{aligned} \nabla G &= \nabla\left(-\frac{1}{4\pi\rho} + \frac{a}{|x_0|}\frac{1}{4\pi\rho^*}\right) \\ &= \frac{x - x_0}{4\pi\rho^3} - \frac{a}{|x_0|}\frac{x - x_0^*}{4\pi\rho^{*3}}. \end{aligned}$$

We recall that $x_0^* = (a/r_0)^2 x_0$. In the case $|x| = a$, we have shown above that $\rho^* = (a/r_0)\rho$. If we put these relations into expression (11.28), we obtain

$$\nabla G = \frac{1}{4\pi\rho^3}\left[x - x_0 - \left(\frac{r_0}{a}\right)^2 x + x_0\right]$$

on the sphere $\partial\Omega$, and thus

$$\frac{\partial G}{\partial n} = \frac{x}{a}\cdot\nabla G = \frac{a^2 - r_0^2}{4\pi a\rho^3}.$$

Now, substituting into the representation formula (11.14), we obtain the solution of problem (11.27) in the form

(11.29) $$\boxed{u(x_0) = \frac{a^2 - |x_0|^2}{4\pi a}\iint\limits_{|x|=a}\frac{h(x)}{|x - x_0|^3}\,\mathrm{d}S.}$$

This is nothing else but the three-dimensional version of the Poisson formula. In the literature, formula (11.29) is often rewritten in spherical coordinates:

$$u(r_0, \theta_0, \phi_0) = \frac{a(a^2 - r_0^2)}{4\pi}\int_0^{2\pi}\int_0^{\pi}\frac{h(\theta, \phi)}{(a^2 + r_0^2 - 2ar_0\cos\psi)^{3/2}}\sin\theta\,\mathrm{d}\theta\,\mathrm{d}\phi,$$

where ψ denotes the angle between the "vectors" x_0 and x.

Remark 11.11 In the same way we can proceed in two dimensions. Let us consider the problem

(11.30)
$$\begin{cases} u_{xx} + u_{yy} = 0, & x^2 + y^2 < a^2, \\ u(x,y) = h(x,y), & x^2 + y^2 = a^2. \end{cases}$$

The corresponding Green's function has the form

$$G(\boldsymbol{x}, \boldsymbol{x}_0) = \frac{1}{2\pi} \ln \rho - \frac{1}{2\pi} \ln\left(\frac{a}{|\boldsymbol{x}_0|}\rho^*\right).$$

The solution of the Dirichlet problem (11.30) is then given by

$$u(\boldsymbol{x}_0) = \frac{a^2 - |\boldsymbol{x}_0|^2}{2\pi a} \int_{|\boldsymbol{x}|=a} \frac{h(\boldsymbol{x})}{|\boldsymbol{x} - \boldsymbol{x}_0|^2} \, ds,$$

which is exactly the Poisson formula (8.17) derived in Chapter 6 in a completely different way.

11.7 Exercises

In the following exercises, $r = \sqrt{x^2 + y^2 + z^2}$ denotes one of the spherical coordinates introduced in Section 11.1.

1. Prove that the solution of the Neumann problem for the Poisson equation is determined in the domain Ω uniquely up to a constant.

2. Prove that Green's function corresponding to the Laplace operator, a domain Ω and a point $\boldsymbol{x}_0 \in \Omega$ is determined uniquely.

3. Prove relation (11.18).

4. Verify formula (11.26) for $G(\boldsymbol{x}, \boldsymbol{0})$.

5. Consider the Dirichlet problem

$$\begin{cases} \Delta u = \lambda u, & \text{in } \Omega, \\ u = 0, & \text{on } \partial\Omega. \end{cases}$$

   Multiply the equation by the function u and integrate it over the domain Ω. Use Green's first identity to prove that a nontrivial solution $u = u(x,y,z)$ can exist only for λ negative.

6. Find radially symmetric solutions of the equation $u_{xx} + u_{yy} + u_{zz} = k^2 u$, where k is a positive constant. Use the substitution $v = u/r$.

$$[u(x,y,z) = \tfrac{1}{r}(Ae^{kr} + Be^{-kr})]$$

Section 11.7 Exercises

7. Solve the equation $u_{xx} + u_{yy} + u_{zz} = 0$ in the shell $\{0 < a < r < b\}$ with the boundary conditions $u = A$ for $r = a$ and $u = B$ for $r = b$, where A and B are constants. Search for a radially symmetric solution.

$$[u(x,y,z) = B + (A - B)(\tfrac{1}{a} - \tfrac{1}{b})^{-1}(\tfrac{1}{r} - \tfrac{1}{b})]$$

8. Solve the equation $u_{xx} + u_{yy} + u_{zz} = 1$ in the shell $\{0 < a < r < b\}$ with the condition $u = 0$ on both the outer and the inner boundary.

$$[u(x,y,z) = \tfrac{1}{6}(r^2 - a^2) - \tfrac{1}{6}ab(a+b)(\tfrac{1}{a} - \tfrac{1}{r})]$$

9. Solve the equation $u_{xx} + u_{yy} + u_{zz} = 1$ in the shell $\{0 < a < r < b\}$ with the conditions $u = 0$ for $r = a$ and $\partial u/\partial r = 0$ for $r = b$. Then consider the limit as $a \to 0$ and give reasons for the result.

10. Show that the homogeneous Robin problem

$$\Delta u = 0 \quad \text{in } \Omega, \qquad \frac{\partial u}{\partial n} + au = 0 \quad \text{on } \partial\Omega,$$

has only the trivial solution $u \equiv 0$. Here Ω is a domain in $\mathbb{R}^3$ and a is a positive constant. Using this result, prove the uniqueness of the boundary value problem

$$\Delta u = f \quad \text{in } \Omega, \qquad \frac{\partial u}{\partial n} + au = g \quad \text{on } \partial\Omega.$$

11. Prove the Dirichlet principle for the Neumann boundary condition. It says that, among *all* real continuously differentiable functions $w(x)$ on Ω, the energy

$$E(w) = \frac{1}{2} \iiint_\Omega |\nabla w|^2 \, dx - \iint_{\partial\Omega} hw \, dS$$

is the smallest for $w = u$, where u solves the Neumann problem

$$-\Delta u = 0 \quad \text{in } \Omega, \qquad \frac{\partial u}{\partial n} = h \quad \text{on } \partial\Omega.$$

It is necessary to assume that the integral average of the given function h is zero. Notice the three main aspects of this principle:

(a) There are *no restrictions* on the function $w(x)$.

(b) The boundary condition $h(x)$ appears in the energy formula.

(c) The energy $E(w)$ remains unchanged if we add an arbitrary constant to the function w.

Proceed in the same way as in the proof of Theorem 11.3.

12. Let $\phi(x)$ be an arbitrary C^2-function defined on $\mathbb{R}^3$ and nonzero outside some ball. Show that

$$\phi(0) = -\frac{1}{4\pi} \iiint \frac{1}{|x|} \Delta\phi(x) \, dx.$$

Here we integrate over the domain where $\phi(x)$ is nonzero.

13. Find Green's function on the half-ball $\Omega = \{x^2 + y^2 + z^2 < a^2,\ z > 0\}$. Consider the solution on the whole ball and use the reflection method similarly to Section 11.6.

    [The result is a sum of four terms involving the distances of x to x_0, x_0^*, $x_0^\#$ and $x_0^{*\#}$, where $*$ denotes reflection across the sphere and $\#$ denotes reflection across the plane $z = 0$.]

14. Find Green's function on the eighth of the ball $\Omega = \{x^2 + y^2 + z^2 < a^2,\ x > 0,\ y > 0,\ z > 0\}$.

15. In the same way as we have defined Green's function on the domain Ω, we can define the so called *Neumann function* $N(x, x_0)$ with the only difference that the property (ii) is replaced by the Neumann boundary condition

    $$\frac{\partial N}{\partial n} = 0 \quad \text{for } x \in \partial\Omega.$$

    Formulate and prove the analogue of Theorem 11.7 on the expression for the solution of the Neumann problem using the Neumann function.

16. Solve the Neumann problem on the half-space:

    $$\begin{cases} \Delta u = 0 & \text{for } z > 0, \\ \frac{\partial u}{\partial z}(x, y, 0) = h(x, y), & u \text{ bounded at infinity}. \end{cases}$$

    Consider the problem for the function $v = \partial u/\partial z$.

    $[u(x, y, z) = C + \int_{-\infty}^{\infty} h(x - \xi) \ln(y^2 + \xi^2)\, d\xi]$

17. Consider the four-dimensional Laplace operator $\Delta u = u_{xx} + u_{yy} + u_{zz} + u_{ww}$. Show that its fundamental symmetric solution is $r^{-3/2}$, where $r^2 = x^2 + y^2 + z^2 + w^2$.

18. Prove the vector form of Green's second identity

    $$\iiint_\Omega (\mathbf{u} \cdot \operatorname{rot} \operatorname{rot} \mathbf{v} - \mathbf{v} \cdot \operatorname{rot} \operatorname{rot} \mathbf{u})\, d\mathbf{x} = \iint_{\partial\Omega} (\mathbf{u} \times \operatorname{rot} \mathbf{v} - \mathbf{v} \times \operatorname{rot} \mathbf{u}) \cdot \mathbf{n}\, dS,$$

    where $\mathbf{u}(x)$, $\mathbf{v}(x)$ are smooth vector-valued functions, Ω is a domain with smooth boundary, $\mathbf{n}$ is the outward normal vector to $\partial\Omega$, and $\mathbf{u} \times \mathbf{v}$ means the vector product of vectors $\mathbf{u}$ and $\mathbf{v}$.

19. Prove Green's first identity for the biharmonic operator Δ^2:

    $$\iiint_\Omega v \Delta^2 u\, d\mathbf{x} = \iiint_\Omega \Delta u \Delta v\, d\mathbf{x} - \iint_{\partial\Omega} \Delta u \frac{\partial v}{\partial n}\, dS + \iint_{\partial\Omega} v \frac{\partial}{\partial n}(\Delta u)\, dS.$$

20. A function u satisfying $\Delta^2 u = 0$ is called biharmonic. Prove the Dirichlet principle for biharmonic functions: "Among all functions v satisfying the boundary conditions
$$v(x) = g(x), \quad \frac{\partial v}{\partial n}(x) = h(x), \quad x \in \partial\Omega,$$
where $g(x)$, $h(x) \in C(\partial\Omega)$, the lowest energy
$$E(v) = \frac{1}{2} \iiint_\Omega |\Delta v|^2 dx$$
is attained by the biharmonic function."

Exercises 6–9, 12–17 are taken from Strauss [19], 10, 11 from Logan [14], and 18–20 from Stavroulakis and Tersian [18].

12 Diffusion Equation in Higher Dimensions

In the previous chapters we considered only one-dimensional models of evolution equations. However, majority of physical phenomena occur in the plane, or in the space. Therefore we now focus on the heat (and diffusion) equation in higher dimensions. We start with its derivation.

12.1 Heat Equation in Three Dimensions

We proceed in a way similar to the one-dimensional case. Let Ω be a domain in $\mathbb{R}^3$ and let $u = u(x, y, z, t)$ denote the temperature at time t and a point $(x, y, z) \in \Omega$. We assume that the material which fills the domain is homogeneous and characterized by a constant mass density ρ and a constant specific heat capacity c. The internal energy at the point (x, y, z) and time t corresponds to the quantity $c\rho u(x, y, z, t)$, the heat flux is a vector function $\boldsymbol{\phi} = \boldsymbol{\phi}(x, y, z, t)$, and the heat sources are described by a scalar function $f = f(x, y, z, t)$. Further, let B be an arbitrary closed ball contained in Ω (see Figure 12.1).

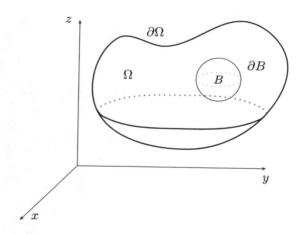

Figure 12.1 *An arbitrary closed ball B in a domain Ω.*

To this ball and an arbitrary time interval $[t_1, t_2]$ we can apply the energy conservation law (in its integral form), which says that the *time change of the internal energy contained in the ball B during the time interval $[t_1, t_2]$ is equal to the sum of the heat flux across the boundary ∂B and the output of the heat sources inside B during $[t_1, t_2]$*.

That is,

$$(12.1) \quad \iiint_B c\rho u(t_2)\,dV - \iiint_B c\rho u(t_1)\,dV = -\int_{t_1}^{t_2}\!\!\iint_{\partial B} \boldsymbol{\phi}\cdot \boldsymbol{n}\,dS\,dt + \int_{t_1}^{t_2}\!\!\iiint_B f\,dV\,dt$$

(cf. the general formula (1.1) or the one-dimensional derivation in Section 5.1). Under some assumptions on continuity and differentiability of the quantities considered and using the Divergence Theorem, we can come to the local (differential) formulation of the conservation law of heat energy:

$$(12.2) \quad c\rho u_t + \operatorname{div}\boldsymbol{\phi} = f$$

for all $(x, y, z) \in \Omega$, $t > 0$.

Now, we use the constitutive law which is, in the case of the heat transfer equation, a three-dimensional version of Fourier's heat law:

$$\boldsymbol{\phi} = -K\operatorname{grad} u.$$

It says that the heat flux is proportional to the temperature gradient with a negative constant of proportionality (the heat flows from warmer places to colder areas); K is the *heat* (or *thermal*) *conductivity*. We put it into (12.2). Moreover, if we realize that

$$\Delta u = \operatorname{div}\operatorname{grad} u = \nabla \cdot \nabla u = u_{xx} + u_{yy} + u_{zz},$$

where Δ denotes the Laplace operator and ∇ the gradient, we obtain the final form of the (nonhomogeneous) heat equation in three dimensions

$$(12.3) \quad \boxed{u_t - k\Delta u = \frac{1}{c\rho}f,}$$

where $k = \frac{K}{c\rho}$ corresponds to the *thermal diffusivity*. This equation describes also the behavior of a diffusing substance in a three-dimensional domain, which is why we call it again a (nonhomogeneous) *diffusion equation*.

Similarly, we would obtain the same expression also in the case of a two-dimensional problem. The Laplace operator has then the form $\Delta u = u_{xx} + u_{yy}$.

12.2 Cauchy Problem in $\mathbb{R}^3$

Homogeneous Problem. Let us consider the Cauchy problem

$$(12.4) \quad \boxed{\begin{aligned} u_t &= k\Delta u, \quad \boldsymbol{x} \in \mathbb{R}^3,\ t > 0, \\ u(\boldsymbol{x}, 0) &= \varphi(\boldsymbol{x}). \end{aligned}}$$

As usual, we denote $\boldsymbol{x} = (x, y, z) \in \mathbb{R}^3$.

As we already know from the one-dimensional case (see Chapter 5), we can express the solution of the Cauchy problem on $\mathbb{R}$ in the integral form

$$u(x,t) = \int_{-\infty}^{\infty} G(x-y, t)\varphi(y)\, dy,$$

where φ is the given initial condition and G the so called *fundamental solution* (diffusion kernel)

$$G(x,t) = \frac{1}{2\sqrt{\pi kt}} e^{-\frac{x^2}{4kt}}.$$

As we will see, the same holds true in higher dimensions.

We start with the following observation. Let $u_1(x,t)$, $u_2(y,t)$, $u_3(z,t)$ be solutions of the one-dimensional diffusion equation. Then $u(x,y,z,t) := u_1(x,t)u_2(y,t)u_3(z,t)$ solves the diffusion equation in $\mathbb{R}^3$. We recommend the reader to verify it by direct substitution. It means that also the function

$$\begin{aligned}G_3(\boldsymbol{x},t) &= G(x,t)G(y,t)G(z,t) \\ &= \frac{1}{8\sqrt{(\pi kt)^3}} e^{-\frac{1}{4kt}(x^2+y^2+z^2)} = \frac{1}{8\sqrt{(\pi kt)^3}} e^{-\frac{1}{4kt}|\boldsymbol{x}|^2}\end{aligned}$$

solves the diffusion equation $u_t - k\Delta u = 0$ in $\mathbb{R}^3$. Since it satisfies

$$\iiint_{\mathbb{R}^3} G_3(\boldsymbol{x},t)\, d\boldsymbol{x} = 1$$

(the reader is kindly asked to verify this fact using the Fubini Theorem), it is again called the *fundamental solution* (or *diffusion kernel*).

Our aim is to show that the solution of the Cauchy problem (12.4) can be again written in the form of a convolution of this fundamental solution $G_3(\boldsymbol{x},t)$ and the initial condition $\varphi(\boldsymbol{x})$, that is,

(12.5) $$u(\boldsymbol{x},t) = \iiint_{\mathbb{R}^3} G_3(\boldsymbol{x}-\boldsymbol{y},t)\varphi(\boldsymbol{y})\, d\boldsymbol{y}.$$

Here $\boldsymbol{y} = (\xi, \eta, \theta) \in \mathbb{R}^3$.

We start with a special initial condition with separated variables

$$\varphi(\boldsymbol{x}) = \phi(x)\psi(y)\zeta(z).$$

In this case we have

$$\begin{aligned}u(\boldsymbol{x},t) &= \iiint_{\mathbb{R}^3} G_3(\boldsymbol{x}-\boldsymbol{y},t)\varphi(\boldsymbol{y})\, d\boldsymbol{y} \\ &= \int_{-\infty}^{\infty} G(x-\xi)\phi(\xi)\, d\xi \int_{-\infty}^{\infty} G(y-\eta)\psi(\eta)\, d\eta \int_{-\infty}^{\infty} G(z-\theta)\zeta(\theta)\, d\theta \\ &= u_1(x,t)\, u_2(y,t)\, u_3(z,t),\end{aligned}$$

where u_1, u_2, u_3 are solutions of the one-dimensional diffusion equation. Thus $u(x,t)$ must solve the three-dimensional diffusion equation. Moreover,

$$\lim_{t \to 0+} u(x,t) = \lim_{t \to 0+} u_1(x,t) \lim_{t \to 0+} u_2(y,t) \lim_{t \to 0+} u_3(z,t)$$
$$= \phi(x)\psi(y)\zeta(z) = \varphi(x)$$

and the initial condition with separated variables is satisfied as well. Due to linearity, the same result must hold true for any finite linear combination of functions with separated variables:

(12.6) $$\varphi(x) = \sum_{k=1}^{n} c_k \phi_k(x) \psi_k(y) \zeta_k(z).$$

It can be shown that any continuous and bounded function on $\mathbb{R}^3$ can be uniformly approximated by functions of type (12.6) on bounded domains. This follows from the properties of Bernstein's polynomials, which go beyond the scope of this book. Nevertheless, this fact is the starting point for the following existence result (see, e.g., Stavroulakis, Tersian [18]).

Theorem 12.1 *Let $\varphi(x)$ be a continuous and bounded function on $\mathbb{R}^3$. Then the solution of the Cauchy problem (12.4) exists and is given by the formula*

$$\boxed{u(x,t) = \iiint_{\mathbb{R}^3} G_3(x-y,t)\varphi(y)\,dy.}$$

Moreover, we have

$$\lim_{t \to 0+} u(x,t) = \varphi(x)$$

uniformly on bounded sets of $\mathbb{R}^3$.

Remark 12.2 The same result holds true in any dimension. In particular, the N-dimensional ($N \geq 1$) fundamental solution assumes the form

$$G_N(x,t) = \frac{1}{2^N \sqrt{(\pi k t)^N}} e^{-\frac{1}{4kt}|x|^2},$$

where $x = (x_1, x_2, \ldots, x_N)$, $|x| = \sqrt{x_1^2 + x_2^2 + \cdots + x_N^2}$, and the solution of the Cauchy problem for a homogeneous diffusion equation on $\mathbb{R}^N$ is given by the formula

$$u(x,t) = \int_{\mathbb{R}^N} G_N(x-y,t)\varphi(y)\,dy.$$

Nonhomogeneous Problem. Using the same approach as in Section 5.3, we can solve a diffusion equation on $\mathbb{R}^3$ with sources:

(12.7)
$$u_t - k\Delta u = f, \quad x \in \mathbb{R}^3, \ t > 0,$$
$$u(x, 0) = \varphi(x).$$

Its solution is given by the formula

$$u(x,t) = \iiint_{\mathbb{R}^3} G_3(x-y,t)\varphi(y)\,dy + \int_0^t \iiint_{\mathbb{R}^3} G_3(x-y, t-s)f(y,s)\,dy\,ds.$$

We let to the reader its derivation and verification.

12.3 Diffusion on Bounded Domains, Fourier Method

In this section we focus on the diffusion equation on a bounded domain and on its solution. That is, we deal with the problem

$$u_t(x,t) = k\Delta u(x,t), \quad x \in \Omega,\ t > 0,$$
$$u(x,t) = h_1(x,t), \quad x \in \Gamma_1,$$
$$\frac{\partial u}{\partial n}(x,t) = h_2(x,t), \quad x \in \Gamma_2,$$
$$\frac{\partial u}{\partial n}(x,t) + au(x,t) = h_3(x,t), \quad x \in \Gamma_3,$$
$$u(x,0) = \varphi(x),$$

where, in general, Ω is a domain in $\mathbb{R}^N$, φ, h_i, $i = 1, 2, 3$ are given functions, a is a given constant, and $\Gamma_1 \cup \Gamma_2 \cup \Gamma_3 = \partial\Omega$. From the physical point of view, such a problem describes the heat flow in a body filling the domain Ω, with the initial temperature given by $\varphi(x)$. On Γ_1 we keep the temperature on the values $h_1(x,t)$; on the boundary Γ_2 we consider the heat flux described by $h_2(x,t)$; and on the third boundary segment Γ_3 we consider the heat exchange with the surrounding medium described by the heat transfer coefficient a and a function $h_3(x,t)$. Usually, $h_3 = aT_0$, where T_0 is the temperature of the surrounding medium.

Similarly, the same problem describes the diffusion process of a gas in the domain Ω. The function $\varphi(x)$ represents the initial concentration. The Dirichlet boundary condition on Γ_1 describes the concentration kept on Γ_1, the Neumann boundary condition on Γ_2 determines the flow of the gas across the boundary, and the Robin boundary condition on Γ_3 describes a certain balance of the gas concentration and its flow across the third boundary segment.

In special cases, some of the boundary segments can be empty.

One way how to solve initial boundary value problems of this type is the application of the *Fourier Method*. We illustrate it in a simpler situation. The following exposition

Section 12.3 Diffusion on Bounded Domains, Fourier Method

is very informative and a lot of stated facts would require deep discussion to make them precise. To make the basic idea clear, we present only formal calculations.

Fourier Method. Let us find a solution of the diffusion equation $u_t = k\Delta u$ on a bounded domain Ω with *homogeneous* Dirichlet, Neumann, or Robin boundary conditions on $\partial\Omega$, and a standard initial condition. The idea of the Fourier method is the same as in the one-dimensional case. First of all, we assume the solution in a separated form
$$u(\boldsymbol{x}, t) = V(\boldsymbol{x})T(t)$$
which, substituted into the equation, results in the identities
$$\frac{T'(t)}{kT(t)} = \frac{\Delta V(\boldsymbol{x})}{V(\boldsymbol{x})} = -\lambda,$$
where λ is a constant. Thus, we come to the *eigenvalue problem* for the Laplace operator

(12.8) $$-\Delta V = \lambda V \quad \text{in } \Omega$$

with general boundary conditions

(12.9) $$V = 0 \quad \text{on} \quad \Gamma_1,$$

(12.10) $$\frac{\partial V}{\partial n} = 0 \quad \text{on} \quad \Gamma_2,$$

(12.11) $$\frac{\partial V}{\partial n} + \alpha V = 0 \quad \text{on} \quad \Gamma_3.$$

It can be shown that the boundary value problem (12.8)–(12.11) has an infinite sequence of nonnegative eigenvalues
$$\lambda_n \to \infty \quad \text{as } n \to \infty$$
and the corresponding complete system of orthogonal eigenfunctions $V_n(\boldsymbol{x})$. The reader should notice similar properties of the eigenvalue problem (12.8)–(12.11) to those of the Sturm-Liouville problem stated in Appendix 14.1. However, it can be very difficult to find the actual values of λ_n.

Returning to the ODE in time variable
$$T'(t) + k\lambda_n T(t) = 0,$$
we obtain a system of time-dependent functions
$$T_n(t) = A_n e^{-k\lambda_n t}.$$
Putting these partial results together, we end up with a solution in the form

(12.12) $$u(\boldsymbol{x}, t) = \sum_{n=1}^{\infty} A_n e^{-k\lambda_n t} V_n(\boldsymbol{x})$$

which satisfies the given initial condition provided the latter is expandable into the Fourier series

$$\varphi(\boldsymbol{x}) = \sum_{n=1}^{\infty} A_n V_n(\boldsymbol{x})$$

according to the system $\{V_n\}_{n=1}^{\infty}$. Using the orthogonality of the eigenfunctions $V_n(\boldsymbol{x})$ (cf. Appendix 14.1), we obtain the formula for the coefficients A_n:

$$A_n = \frac{\int_\Omega \varphi(\boldsymbol{x}) V_n(\boldsymbol{x}) \mathrm{d}\boldsymbol{x}}{\int_\Omega V_n^2(\boldsymbol{x}) \mathrm{d}\boldsymbol{x}}.$$

(Since Ω is, in general, a domain in $\mathbb{R}^N$, the integrals above are also N-dimensional!)

We illustrate the previous steps by a concrete example.

Example 12.3 Let us find the temperature distribution in the rectangle $\Omega = (0, a) \times (0, b)$ whose boundary is kept on zero temperature; the initial distribution is given by a function $\varphi(x, y)$. That is, we solve the initial boundary value problem

(12.13)
$$\begin{cases} u_t = k(u_{xx} + u_{yy}), & (x, y) \in (0, a) \times (0, b),\ t > 0, \\ u(0, y, t) = u(a, y, t) = u(x, 0, t) = u(x, b, t) = 0, \\ u(x, y, 0) = \varphi(x, y). \end{cases}$$

First of all, we again separate the time and space variables:

$$u(x, y, t) = V(x, y) T(t)$$

thus obtaining the system of equations

$$\frac{T'(t)}{kT(t)} = \frac{V_{xx}(x, y) + V_{yy}(x, y)}{V(x, y)} = -\lambda,$$

where λ is a constant. This yields

$$T' + k\lambda T = 0, \quad V_{xx} + V_{yy} + \lambda V = 0.$$

Now, we will have a look at the spatial problem in more detail. Since it is a linear stationary PDE in a rectangle, we can use the separation of variables again. Thus, we look for its solution in the form

$$V(x, y) = X(x) Y(y).$$

After substitution, we obtain

$$X''Y + XY'' + \lambda XY = 0$$

and hence, dividing by XY,

$$\frac{X''(x)}{X(x)} = -\frac{Y''(y)}{Y(y)} - \lambda.$$

Section 12.3 Diffusion on Bounded Domains, Fourier Method

Since the left-hand side depends only on the x-variable and the right-hand side depends only on the y-variable, we conclude that both sides of the identity must be equal to a constant, say, $-\mu$. Thus, we obtain two separated ODEs

$$X'' + \mu X = 0, \quad Y'' + \nu Y = 0,$$

where $\nu := \lambda - \mu$. To satisfy the homogeneous boundary conditions in (12.13), the functions X and Y must satisfy the conditions

$$X(0) = X(a) = 0, \quad Y(0) = Y(b) = 0.$$

Starting with the problem in the x-variable, we obtain

$$\mu_n = \frac{n^2 \pi^2}{a^2}, \quad X_n(x) = \sin \frac{n \pi x}{a}, \quad n \in \mathbb{N}.$$

Similarly, for the problem in the y-variable, we get

$$\nu_m = \frac{m^2 \pi^2}{b^2}, \quad Y_m(x) = \sin \frac{m \pi y}{b}, \quad m \in \mathbb{N}.$$

Thus, the eigenvalues of the Laplace operator form a sequence

$$\lambda_{mn} = \mu_n + \nu_m = \frac{n^2 \pi^2}{a^2} + \frac{m^2 \pi^2}{b^2}, \quad m, n \in \mathbb{N},$$

and the corresponding eigenfunctions

$$V_{mn}(x,y) = X_n(x) Y_m(y) = \sin \frac{n \pi x}{a} \sin \frac{m \pi y}{b}, \quad m, n \in \mathbb{N},$$

form an orthogonal system on $(0, a) \times (0, b)$. (Notice the double index!) It can be proved that this system is complete and hence λ_{mn} describe the set of all eigenvalues of the Laplace operator on $(0, a) \times (0, b)$ with homogeneous Dirichlet boundary conditions. Now, we solve the time problems $T' + k \lambda_{mn} T = 0$, that is,

$$T_{mn} = A_{mn} e^{-k \lambda_{mn} t}.$$

Hence, the solution of the original two-dimensional diffusion equation can be written as

(12.14) $$u(x, y, t) = \sum_{m=1}^{\infty} \sum_{n=1}^{\infty} A_{mn} e^{-k \lambda_{mn} t} \sin \frac{n \pi x}{a} \sin \frac{m \pi y}{b}.$$

Functions of this form already satisfy homogeneous Dirichlet boundary conditions. The remaining unused information is the initial condition. We can conclude that for all initial conditions which are expandable into a series

(12.15) $$\varphi(x, y) = \sum_{m=1}^{\infty} \sum_{n=1}^{\infty} A_{mn} \sin \frac{n \pi x}{a} \sin \frac{m \pi y}{b},$$

the solution of problem (12.13) is given by formula (12.14). Using the orthogonality of the eigenfunctions, we can determine the constants A_{mn}:

$$A_{mn} = \frac{\int_0^a \int_0^b \varphi(x,y) V_{mn}(x,y)\, dy\, dx}{\int_0^a \int_0^b V_{mn}^2(x,y)\, dy\, dx}.$$

$\square$

Figure 12.2 depicts the solution of problem (12.13) with the constant initial condition $\varphi(x) = 100$ and with the data $a = 1$, $b = 1$, $k = 1$. The first graph corresponds to the approximated initial condition, the other three graphs illustrate the approximated solution at times $t = 0.01$, $t = 0.04$ and $t = 0.09$. We used formulae (12.14) and (12.15) with partial sums up to $m = 15$, $n = 15$.

The same approach can be used also for other types of boundary conditions and for similar problems in higher dimensions.

Example 12.4 We solve the initial boundary value problem for the diffusion equation $u_t = k\Delta u$ in the cube $\Omega = \{0 < x < \pi,\ 0 < y < \pi,\ 0 < z < \pi\}$. This time, consider homogeneous Neumann boundary conditions

$$u_x(0, y, z, t) = u_x(\pi, y, z, t) = 0,$$
$$u_y(x, 0, z, t) = u_y(x, \pi, z, t) = 0,$$
$$u_z(x, y, 0, t) = u_z(x, y, \pi, t) = 0,$$

and initial condition

$$u(x, y, z, 0) = \varphi(x, y, z).$$

We proceed in the same way as in the previous example. First, we separate the time and space variables to obtain

$$T' + k\lambda T = 0, \quad V_{xx} + V_{yy} + V_{zz} + \lambda V = 0.$$

Second, we apply the separation of variables to the spatial problem. Thus, we get three ODEs

$$\begin{aligned} X'' + \mu X &= 0, \\ Y'' + \nu Y &= 0, \\ Z'' + \eta Z &= 0, \end{aligned}$$

where $\lambda = \mu + \nu + \eta$. Adding the boundary conditions $X'(0) = X'(\pi) = Y'(0) = Y'(\pi) = Z'(0) = Z'(\pi) = 0$ and solving the corresponding ordinary problems, we obtain

$$\mu_n = n^2, \quad \nu_m = m^2, \quad \eta_l = l^2, \quad l, m, n \in \mathbb{N} \cup \{0\}$$

and

$$X_n(x) = \cos nx, \quad Y_m(y) = \cos my, \quad Z_l(z) = \cos lz, \quad l, m, n \in \mathbb{N} \cup \{0\}.$$

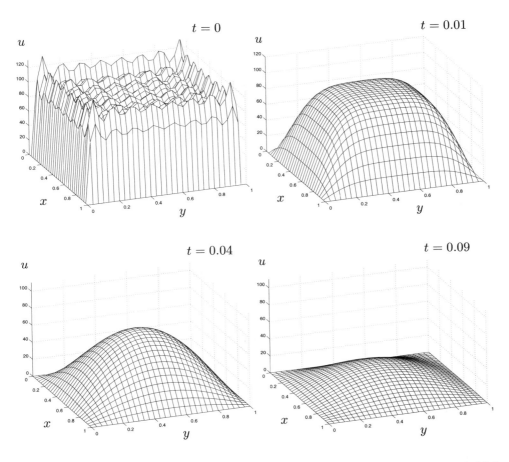

Figure 12.2 *Solution of the initial boundary value problem (12.13) with constant initial condition on time levels $t = 0$, 0.01, 0.04, 0.09.*

Thus, the eigenvalues of the Laplace operator in the cube $\Omega = \{0 < x < \pi,\ 0 < y < \pi,\ 0 < z < \pi\}$ with homogeneous Neumann boundary conditions form a sequence

$$\lambda_{lmn} = m^2 + n^2 + l^2, \quad m, n, l \in \mathbb{N} \cup \{0\},$$

with the corresponding system of eigenfunctions

$$V_{lmn} = \cos nx \cos my \cos lz, \quad m, n, l \in \mathbb{N} \cup \{0\}.$$

Notice that in this case the problem has zero eigenvalue with a constant eigenfunction.

Now, we can continue with the time problem. In the same way as in the previous cases, we obtain

$$T_{lmn}(t) = B_{lmn} e^{-k\lambda_{lmn} t} \quad \text{for } (m, n, l) \neq (0, 0, 0)$$

and
$$T_{000}(t) = B_{000},$$
and we can conclude that the required solution is given by
$$u(x,y,z,t) = \sum_{l=0}^{\infty}\sum_{m=0}^{\infty}\sum_{n=0}^{\infty} B_{lmn} e^{-k\lambda_{lmn} t} \cos nx \cos my \cos lz,$$
where the coefficients B_{lmn} follow from the expansion of the initial condition
$$\varphi(x,y,z) = \sum_{l=0}^{\infty}\sum_{m=0}^{\infty}\sum_{n=0}^{\infty} B_{lmn} \cos nx \cos my \cos lz.$$
That is,

(12.16) $$B_{lmn} = \frac{2^3}{\pi^3} \int_0^\pi \int_0^\pi \int_0^\pi \varphi(x,y,z) \cos nx \cos my \cos lz \, dx\, dy\, dz$$

for $n,m,l > 0$; for B_{0mn}, B_{l0n}, or B_{lm0} we have to use one-half of (12.16), for B_{00n}, B_{0m0}, B_{l00} we use one-fourth, and for B_{000} we use one-eighth of (12.16). $\square$

As we can see, the geometry (in particular, the rectangularity) of the domain Ω is crucial for easy determination of the eigenvalues and the corresponding eigenfunctions of the problems considered. We already know that other domains which allow the application of the Fourier method, are a disc and a ball (or their suitable parts), since they both become rectangular under the transformation into the polar or spherical coordinates, respectively. Moreover, in the radially symmetric situations, the problems are considerably simplified.

Example 12.5 (Diffusion in Disc.) Let us consider the heat problem in the disc

(12.17) $$\begin{cases} u_t = k\Delta u, & x^2 + y^2 < a^2,\ t > 0, \\ u(x,y,t) = 0, & x^2 + y^2 = a^2,\ t > 0, \\ u(x,y,0) = \varphi(\sqrt{x^2+y^2}), & x^2+y^2 < a^2. \end{cases}$$

Since the domain is circular, we will transform the problem into polar coordinates (r,θ). Moreover, the problem data (that is, the boundary and initial conditions) do not depend on the angle θ, thus the solution u can be expected to be radially symmetric and we solve the simplified problem in two variables r and t:

$$\begin{cases} u_t = k(u_{rr} + \frac{1}{r}u_r), & 0 < r < a,\ t > 0, \\ u(r,t) = 0, & r = a,\ t > 0, \\ u(r,0) = \varphi(r), & 0 \le r < a. \end{cases}$$

As usual, we separate the variables
$$u(r,t) = R(r)T(t)$$

Section 12.3 Diffusion on Bounded Domains, Fourier Method

and obtain
$$\frac{T'(t)}{kT(t)} = \frac{R''(r) + \frac{1}{r}R'(r)}{R(r)} = -\lambda.$$

The spatial ODE is the so called *Bessel equation*
$$R''(r) + \frac{1}{r}R'(r) + \lambda R(r) = 0$$

which has a pair of linearly independent solutions. The first, which is finite at $r=0$, is the *Bessel function of order zero*
$$R(r) = J_0(\sqrt{\lambda}r) = \sum_{j=0}^{\infty}(-1)^j \frac{(\sqrt{\lambda}r/2)^{2j}}{(j!)^2}.$$

The second solution of the Bessel equation is infinite at $r = 0$ and thus we are not interested in it. (For more details, see Appendix 14.2.) Further, we have to satisfy the homogeneous boundary condition on the boundary $r = a$, that is,
$$R(a) = J_0(\sqrt{\lambda}a) = 0.$$

Thus we get a sequence of eigenvalues
$$\lambda_n = \frac{1}{a^2}\mu_n^2, \quad n \in \mathbb{N},$$

and corresponding eigenfunctions
$$R_n(r) = J_0(\sqrt{\lambda_n}r) = J_0(\mu_n \frac{r}{a}), \quad n \in \mathbb{N}.$$

Here μ_n are the roots of the Bessel function J_0. (Each Bessel function has an infinite number of positive roots that go to infinity, cf. Appendix 14.2.)

Now, we go back to the time problem, which has the standard solution
$$T_n(t) = A_n e^{-k\lambda_n t}.$$

The solution of the original problem (12.17) then can be written in the form
$$u(r,t) = \sum_{n=1}^{\infty} A_n e^{-k\lambda_n t} J_0(\sqrt{\lambda_n}r)$$

and satisfies the initial condition
$$\varphi(r) = \sum_{n=1}^{\infty} A_n J_0(\sqrt{\lambda_n}r).$$

For $\rho = \frac{r}{a} \in [0,1]$ we have
$$\varphi(a\rho) = \sum_{n=1}^{\infty} A_n J_0(\mu_n \rho)$$

and the properties of the Bessel functions stated at the end of Appendix 14.2 imply

$$A_n = \frac{2}{J_0'^2(\mu_n)} \int_0^1 \rho J_0(\mu_n \rho) \varphi(a\rho) \, d\rho$$

(the reader is kindly asked to carry out detailed calculations).

□

Figure 12.3 depicts the solution of problem (12.17) for the choice $a = 1$, $k = 1$ and with the initial condition

(12.18) $$\varphi(x, y) = \varphi(r) = J_0(\mu_1 r) + J_0(\mu_2 r),$$

where J_0 is the Bessel function of order zero and μ_1, μ_2 are its first two roots. Notice that, for this data, the solution assumes the form

$$u(x, y, t) = u(r, t) = e^{-\mu_1^2 t} J_0(\mu_1 r) + e^{-\mu_2^2 t} J_0(\mu_2 r).$$

The particular graphs in Figure 12.3 correspond to the solution at times $t = 0$, $t = 0.01$, $t = 0.04$ and $t = 0.09$.

Remark 12.6 (Nonhomogeneous Problems.) The idea of solving nonhomogeneous problems for the diffusion equation in higher dimensions is completely the same as in the one-dimensional case. If we solve a *nonhomogeneous equation*, we find the system of eigenfunctions $V_n(x)$ corresponding to the homogeneous problem and expand all the problem data (that is, the right-hand side, the initial condition, as well as the searched solution) to Fourier series with respect to this system. Using its completeness and orthogonality, we split the original PDE problem into an infinite system of ODEs in the time variable which are easy to solve.

Problems with *nonhomogeneous boundary conditions* can cause more trouble. The idea is, again, to split the solution into two parts: the first part corresponds to the solution satisfying the equation with homogeneous boundary conditions, while the second "stationary" (or "quasi-stationary") part respects the nonhomogeneous boundary conditions. In one-dimensional cases, we usually "guessed" the stationary part easily (see Section 7.3). In higher dimensions, it means to solve the Laplace equation with given nonhomogeneous boundary conditions, which can be very laborious. Again, the rectangularity of the domain can be essential.

Remark 12.7 (General Principles for Diffusion Equation.) The aim of this remark is to recall all basic properties of the diffusion equation which remain unchanged in any dimension.

- First of all, the solution formula in Theorem 12.1 implies that diffusion (as well as heat) propagates at *infinite speed*. (After any short time, the solution is nonzero

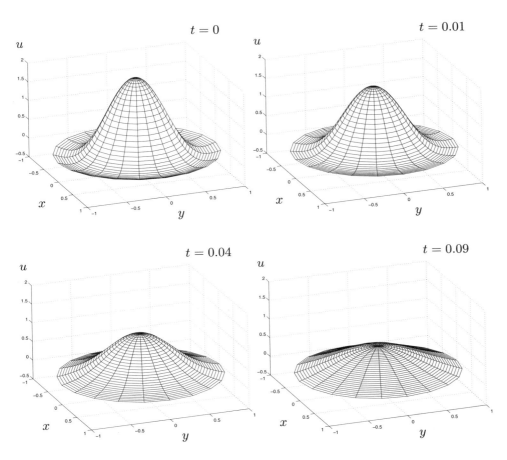

Figure 12.3 *Solution of the initial boundary value problem (12.17) with initial condition (12.18) on time levels $t = 0$, 0.01, 0.04, 0.09.*

everywhere, even if the initial condition was nonzero only on a small domain.) As we have already mentioned in Chapter 5, this fact reflects the inaccuracy of the diffusion model. However, the incurred error is very small and the diffusion equation can be used as a good approximation of many real problems.

- Another property which occurs in any dimension is the *ill-posedness* of the diffusion problems for $t < 0$. It is not possible to determine the temperature of the body backwards in time, neither to find the original concentration of a diffusing gas, provided we know only the actual state.

- A very important property of the diffusion equation on any (bounded or unbounded) domain in any dimension is the *Maximum Principle*. Its strong version says that the maximum and minimum values of the solution are achieved only on the bottom or jacket of the space-time cylinder, unless the solution is constant. Here the bottom of the cylinder is the (in general, N-dimensional) domain Ω at time $t = 0$, and the jacket represents the boundary $\partial\Omega$ at any time $t > 0$! See Figure 12.4.

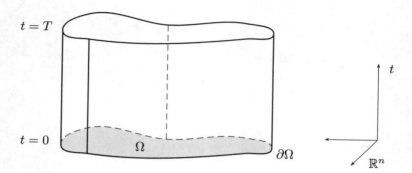

Figure 12.4 *Space-time cylinder $\Omega \times [0, T]$.*

- As in the one-dimensional case, the Maximum Principle has many consequences. The most important ones are the *uniqueness* and *uniform stability* of the solution. Studying the corresponding proofs in Section 10.4, notice their independence of the dimension.

- The uniqueness and stability properties can be obtained also via the *energy method*, which can be applied in any dimension. We again refer to Section 10.5.

12.4 Exercises

1. Solve the following problem for the diffusion equation on the whole space:
$$\begin{cases} u_t - \Delta u = 0, & (x, y, z) \in \mathbb{R}^3, \ t > 0, \\ u(x, y, z, 0) = x^2 yz. \end{cases}$$

$$[u(x, y, z, t) = x^2 yz + 2tyz]$$

2. Solve the following problem for the diffusion equation on the whole space:
$$\begin{cases} u_t - \Delta u = 0, & (x, y, z) \in \mathbb{R}^3, \ t > 0, \\ u(x, y, z, 0) = x^2 yz - xyz^2. \end{cases}$$

$$[u(x, y, z, t) = y(xz - 2t)(x - z)]$$

Section 12.4 Exercises

3. Using the reflection method (method of odd extension), find a formula for the solution of the initial boundary value problem for the diffusion equation in the half-plane
$$\begin{cases} u_t - k\Delta u = 0, & x > 0,\ y \in \mathbb{R},\ t > 0, \\ u(0, y, t) = 0, \\ u(x, y, 0) = \varphi(x, y). \end{cases}$$
$[u(x, y, t) = \int_{-\infty}^{\infty} \int_{0}^{\infty} (G_2(x - \xi, y - \eta, t) - G_2(x + \xi, y - \eta, t))\varphi(\xi, \eta)\, d\xi\, d\eta]$

4. Using the reflection method (method of even extension), find a formula for the solution of the initial boundary value problem for the diffusion equation in the half-space
$$\begin{cases} u_t - k\Delta u = 0, & (x, y) \in \mathbb{R}^2,\ z > 0,\ t > 0, \\ u_z(x, y, 0, t) = 0, \\ u(x, y, z, 0) = \varphi(x, y, z). \end{cases}$$
$[u(x, y, z, t) = \int_{-\infty}^{\infty} \int_{-\infty}^{\infty} \int_{0}^{\infty} (G_3(x - \xi, y - \eta, z - \theta, t) - G_3(x + \xi, y - \eta, z - \theta, t))\varphi(\xi, \eta, \theta)\, d\xi\, d\eta\, d\theta]$

5. Solve the diffusion equation $u_t = u_{xx} + u_{yy}$ in the disc $x^2 + y^2 < 1$ with homogeneous Dirichlet boundary condition and with the initial condition $u(x, y, 0) = 1 - x^2 - y^2$.

[in polar coordinates: $u(r, t) = 8\sum_{k=1}^{\infty} e^{-\mu_k^2 t} \frac{J_0(\mu_k r)}{\mu_k^3 J_1(\mu_k)}$.]

6. Solve the problem
$$\begin{cases} u_t = a^2(u_{rr} + \frac{1}{r}u_r), & 0 < r < R,\ t > 0, \\ u(r, 0) = T, \\ \frac{\partial}{\partial r} u(r, t)|_{r=R} = q. \end{cases}$$

$[u(r, t) = T + qR(2\frac{a^2 t}{R^2} - \frac{1}{4}(1 - 2\frac{r^2}{R^2})) - \sum_{n=1}^{\infty} \frac{2e^{-(a\mu_n/R)^2 t}}{\mu_n^2 J_0(\mu_n)} J_0(\frac{\mu_n r}{R})$, where μ_n are positive roots of J_1]

7. Consider the problem of cooling of the ball of radius R with a radiation boundary condition
$$u_r(R, t) = -hu(R, t),$$
where h is a positive constant and $Rh < 1$. Assume that the initial temperature $u(x, t) = \varphi(r)$ depends only on the radius r. Solve the radially symmetric diffusion equation using the Fourier method. (The eigenvalues λ_n are obtained as the positive roots of the equation $\tan R\lambda = \frac{R\lambda}{1 - Rh}$.)

8. Consider a thin rectangular plate of length a and width b with perfect lateral insulation. Find the distribution of temperature in the plate for the following data: $a = 2\pi$, $b = 4\pi$, $k = 1$, boundary conditions
$$u_x(0, y, t) = 0, \quad u_x(a, y, t) = 0,$$
$$u_y(x, 0, t) = 0, \quad u_y(x, b, t) = 0,$$

and the initial condition

$$u(x,y,0) = \cos 3x, \quad 0 \le x \le a, \; 0 \le y \le b.$$

9. Find the distribution of temperature in a semicircular plate ($0 < \theta < \pi$) of radius 1 with the initial condition $u(r,\theta,0) = g(r,\theta)$ and the boundary conditions

   (a) $u(r,0,t) = 0, \quad u(r,\pi,t) = 0, \quad u(1,\theta,t) = 0.$
   (b) $u_\theta(r,0,t) = 0, \quad u_\theta(r,\pi,t) = 0, \quad u(1,\theta,t) = 0.$

   What occurs after infinitely long time?

Exercises 1–6 are taken from Stavroulakis and Tersian [18], 7 from Logan [14], 8–9 from Keane [12].

13 Wave Equation in Higher Dimensions

13.1 Membrane Vibrations and Wave Equation in Two Dimensions

The two-dimensional analogue of the oscillating string is a vibrating membrane fastened on a fixed frame. Let us consider only vertical oscillations and denote the displacement at a point (x, y) and time t by $u(x, y, t)$. Let us choose an arbitrary fixed subdomain Ω of the membrane (see Figure 13.1) and apply Newton's Second Law of Motion and the mass conservation law. We proceed in a way similar to that used in one dimension (see Section 4.1).

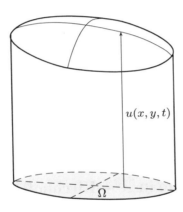

Figure 13.1 *Vibrating membrane over the subdomain Ω.*

The application of both the above mentioned conservation laws leads to the equality

(13.1) $$\iint_\Omega \rho_0(x,y) u_{tt}(x,y,t) \, dx \, dy = \int_{\partial \Omega} T(x,y,t) \frac{\partial u}{\partial n} \, ds$$

(cf. relation (4.6) for the one-dimensional string). Here $\partial \Omega$ denotes the boundary of the domain Ω and we assume that the normal exists at every point of $\partial \Omega$. The function $T = T(x, y, t)$ represents the inner tension of the membrane and $\rho_0 = \rho_0(x, y)$ denotes the mass density distribution of the membrane at time $t = 0$. Since we do not consider any motion in the horizontal direction, we can assume $T(x, y, t)$ independent of x and y, that is, $T(x, y, t) = \tau(t)$. If we use Green's Theorem, we can transform the curve integral on the right-hand side of equation (13.1) to the surface integral:

$$\int_{\partial\Omega} \tau(t)\frac{\partial u}{\partial n}\,ds = \iint_{\Omega} \operatorname{div}(\tau(t)\operatorname{grad} u)\,dx\,dy.$$

Consequently,

$$\iint_{\Omega} \rho_0(x,y) u_{tt}(x,y,t)\,dx\,dy = \iint_{\Omega} \tau(t)\underbrace{\operatorname{div}(\operatorname{grad} u)}_{\Delta u}\,dx\,dy,$$

where $\Delta u = u_{xx} + u_{yy}$ is the Laplace operator in two dimensions. Since the domain Ω has been chosen arbitrarily, we can come (imposing some smoothness assumptions on u) to the differential (local) formulation. Moreover, if we consider the mass density ρ_0 as well as the tension τ to be constant, that is, $\rho_0(x,y) = \rho_0$, $\tau(t) = \tau_0$, and if we denote $c = \sqrt{\tau_0/\rho_0}$, we obtain the relation

(13.2) $$\boxed{u_{tt} = c^2 \Delta u.}$$

It is obvious that a similar process leads to a similar relation in the three-dimensional case, where, however, $\Delta u = u_{xx} + u_{yy} + u_{zz}$ is the Laplace operator in three dimensions. From the physical point of view, models of this type can describe vibrations in an elastic body, propagation of sound waves in the air, propagation of seismic waves in the earth crust, electromagnetic waves, etc.

If u in (13.2) is an unknown function, we speak about the wave equation in two or three spatial variables, respectively.

13.2 Cauchy Problem in $\mathbb{R}^3$—Kirchhoff's Formula

Let us consider the Cauchy problem for the wave equation in $\mathbb{R}^3$

(13.3) $$\boxed{\begin{aligned}&u_{tt} = c^2 \Delta u, \quad \boldsymbol{x} = (x,y,z) \in \mathbb{R}^3,\ t > 0,\\ &u(\boldsymbol{x},0) = \varphi(\boldsymbol{x}), \quad u_t(\boldsymbol{x},0) = \psi(\boldsymbol{x}).\end{aligned}}$$

We start with finding the explicit formula for its solution.

Theorem 13.1 (Kirchhoff's Formula.) *Let $\varphi \in C^3(\mathbb{R}^3)$ and $\psi \in C^2(\mathbb{R}^3)$. The classical solution of the Cauchy problem for the homogeneous wave equation (13.3) exists, it is unique and is given by the formula*

(13.4) $$\boxed{u(\boldsymbol{x}_0, t) = \frac{1}{4\pi c^2 t}\iint_{|\boldsymbol{x}-\boldsymbol{x}_0|=ct} \psi(\boldsymbol{x})\,dS + \frac{\partial}{\partial t}\left(\frac{1}{4\pi c^2 t}\iint_{|\boldsymbol{x}-\boldsymbol{x}_0|=ct} \varphi(\boldsymbol{x})\,dS\right).}$$

Section 13.2 Cauchy Problem in $\mathbb{R}^3$—Kirchhoff's Formula

Here the integrals are surface integrals over the sphere with its center at x_0 and radius ct. This formula is known as *Kirchhoff's formula* but its author is Poisson. For its derivation, we use the so called *spherical means*.

Let us denote by $\bar{u}(x_0, r, t)$ the mean (average) value of the function $u(x, t)$ over the sphere $|x - x_0| = r$, that is

$$\bar{u}(x_0, r, t) = \frac{1}{4\pi r^2} \iint\limits_{|x-x_0|=r} u(x, t) \mathrm{d}S.$$

Using transformation to spherical coordinates, we can write

$$\bar{u}(x_0, r, t) = \frac{1}{4\pi} \int_0^{2\pi} \int_0^{\pi} u(r, \theta, \varphi, t) \sin\varphi \, \mathrm{d}\varphi \, \mathrm{d}\theta,$$

where

$$u(r, \theta, \varphi, t) = u(x_0 + r\cos\theta \sin\varphi, y_0 + r\sin\theta \sin\varphi, z_0 + r\cos\varphi, t).$$

Proof (Derivation of Kirchhoff's Formula.) The main idea of the derivation of Kirchhoff's formula consists in two steps. First, we solve the problem (13.3) for the spherical means, and, second, we pass to the solution of the original problem using the relation

(13.5) $$u(x_0, t) = \lim_{r \to 0} \bar{u}(x_0, r, t).$$

Let us start with the following observation: if u satisfies the wave equation, then $\bar{u}$ satisfies it as well. Indeed, the equality $\overline{u_{tt}} = (\bar{u})_{tt}$ is obvious. Further, using spherical coordinates and the rotational invariance of the Laplace operator, we obtain

(13.6) $$\overline{\Delta u} = \Delta \bar{u} = \bar{u}_{rr} + \frac{2}{r} \bar{u}_r.$$

(The direct derivation of (13.6) is required in Exercise 3 in Section 13.7.) Thus, $\bar{u}$ satisfies the equation

(13.7) $$\boxed{\bar{u}_{tt} = c^2 \left(\bar{u}_{rr} + \frac{2}{r} \bar{u}_r \right).}$$

Now, we introduce the substitution

(13.8) $$v(r, t) = r\bar{u}(x_0, r, t).$$

Since $v_{tt} = r\bar{u}_{tt}$, $v_r = r\bar{u}_r + \bar{u}$ and $v_{rr} = r\bar{u}_{rr} + 2\bar{u}_r$, equation (13.7) reduces to

(13.9) $$\boxed{v_{tt} = c^2 v_{rr}}$$

for $(r, t) \in (0, \infty) \times (0, \infty)$. Obviously, we can set

(13.10) $$v(0, t) = 0.$$

Moreover, since u solves the original Cauchy problem (13.3), v must fulfil the initial conditions

(13.11) $\qquad v(r,0) = r\overline{\varphi}(\boldsymbol{x}_0, r), \qquad v_t(r,0) = r\overline{\psi}(\boldsymbol{x}_0, r).$

However, equation (13.9) with the boundary condition (13.10) and initial conditions (13.11) forms a standard one-dimensional problem for the wave equation on the half-line. Its solution was found in Section 7.1 and for $0 \le r \le ct$ it can be written in the form

$$v(r,t) = \frac{1}{2}[(ct+r)\overline{\varphi}(\boldsymbol{x}_0, ct+r) - (ct-r)\overline{\varphi}(\boldsymbol{x}_0, ct-r)] + \frac{1}{2c}\int_{ct-r}^{ct+r} s\overline{\psi}(\boldsymbol{x}_0, s)\,\mathrm{d}s.$$

If we rewrite the first term on the right-hand side, we obtain an equivalent formula

(13.12) $\qquad v(r,t) = \dfrac{1}{2c}\left(\dfrac{\partial}{\partial t}\displaystyle\int_{ct-r}^{ct+r} s\overline{\varphi}(\boldsymbol{x}_0,s)\,\mathrm{d}s + \displaystyle\int_{ct-r}^{ct+r} s\overline{\psi}(\boldsymbol{x}_0,s)\,\mathrm{d}s\right)$

for $0 \le r \le ct$.

Now we determine the value of $u(\boldsymbol{x}_0, t)$. As we have stated above, we use relation (13.5), that is,

$$\begin{aligned} u(\boldsymbol{x}_0, t) &= \lim_{r \to 0} \overline{u}(\boldsymbol{x}_0, r, t) = \lim_{r \to 0} \frac{v(r,t)}{r} \\ &= \lim_{r \to 0} \frac{v(r,t) - v(0,t)}{r} = \frac{\partial v}{\partial r}(0,t). \end{aligned}$$

We differentiate (13.12) to obtain

$$\begin{aligned} \frac{\partial v}{\partial r} &= \frac{1}{2c}\frac{\partial}{\partial t}[(ct+r)\overline{\varphi}(\boldsymbol{x}_0, ct+r) + (ct-r)\overline{\varphi}(\boldsymbol{x}_0, ct-r)] \\ &\quad + \frac{1}{2c}[(ct+r)\overline{\psi}(\boldsymbol{x}_0, ct+r) + (ct-r)\overline{\psi}(\boldsymbol{x}_0, ct-r)]. \end{aligned}$$

Putting $r = 0$, we get

$$\begin{aligned} u(\boldsymbol{x}_0, t) &= \frac{\partial v}{\partial r}(0,t) = \frac{1}{2c}\frac{\partial}{\partial t}\left((2ct)\overline{\varphi}(\boldsymbol{x}_0, ct)\right) + \frac{1}{2c}(2ct)\overline{\psi}(\boldsymbol{x}_0, ct) \\ &= \frac{\partial}{\partial t}\left(t\overline{\varphi}(\boldsymbol{x}_0, ct)\right) + t\overline{\psi}(\boldsymbol{x}_0, ct) \\ &= \frac{\partial}{\partial t}\left(\frac{1}{4\pi c^2 t}\iint_{|\boldsymbol{x}-\boldsymbol{x}_0|=ct} \varphi(\boldsymbol{x})\,\mathrm{d}S\right) + \frac{1}{4\pi c^2 t}\iint_{|\boldsymbol{x}-\boldsymbol{x}_0|=ct} \psi(\boldsymbol{x})\,\mathrm{d}S, \end{aligned}$$

which is exactly Kirchhoff's formula (13.4).

The uniqueness of the classical solution is a consequence of the linearity of the equation and can be proved easily (see Exercise 13 in Section 13.7). ∎

Remark 13.2 Unlike the one-dimensional case when the solution given by d'Alembert's formula is as regular as the initial displacement, here the solution is less regular because of the time derivative in Kirchhoff's formula. In general, if $\varphi \in C^{n+1}(\mathbb{R}^3)$ and $\psi \in C^n(\mathbb{R}^3)$, $n \geq 2$, then u is of the class C^n on $\mathbb{R}^3 \times (0, \infty)$. If φ and ψ are both of class C^2, then the second derivatives of u can be unbounded at some points and the solution is not the classical one. This fact is known as the *focusing effect*.

Huygens' principle. Let us notice that, according to Kirchhoff's formula, the solution of (13.3) at the point (x_0, t) depends only on the values of $\varphi(x)$ and $\psi(x)$ for x from the *spherical surface* $|x - x_0| = ct$, but it does not depend on the values of the initial data *inside* this sphere. Similarly, using the opposite point of view we conclude that the values of φ and ψ at a point $x_1 \in \mathbb{R}^3$ influence the solution of the three-dimensional wave equation only on the *spherical surface* $|x - x_1| = ct$. This phenomenon is called *Huygens' principle*.

This principle corresponds to the fact that, in the "three-dimensional world", solutions of the wave equation propagate *exactly* at the speed c. For instance, any electromagnetic signal in a vacuum propagates exactly at the speed of light, or any sound is carried through the air exactly at the speed of sound without any "echoes" (assuming no barriers). This means that the listener hears at time t what the speaker said exactly at time $(t - d/c)$ (here d is the distance between the persons), and not a mess of sounds produced at different times.

As we already know from d'Alembert's formula, this principle does not hold true in one dimension, and as we shall see later, neither in two dimensions.

13.3 Cauchy problem in $\mathbb{R}^2$

Let us consider now the Cauchy problem for the homogeneous wave equation in $\mathbb{R}^2$

(13.13) $$\begin{array}{l} u_{tt} = c^2(u_{xx} + u_{yy}), \quad (x, y) \in \mathbb{R}^2, \ t > 0, \\ u(x, y, 0) = \varphi(x, y), \quad u_t(x, y, 0) = \psi(x, y). \end{array}$$

We can handle it as a "special three-dimensional problem" the solution of which does not depend on the variable z. Then, according to Kirchhoff's formula, the solution $u = u(x_0, t) = u(x_0, y_0, 0, t)$ satisfies

(13.14) $$u(x_0, t) = \frac{1}{4\pi c^2 t} \iint_{|x-x_0|=ct} \psi(x) \, dS + \frac{\partial}{\partial t}\left(\frac{1}{4\pi c^2 t} \iint_{|x-x_0|=ct} \varphi(x) \, dS \right).$$

Here $x_0 = (x_0, y_0, 0)$, $x = (x, y, z)$ and $\varphi(x) := \varphi(x, y)$, $\psi(x) := \psi(x, y)$ for any z. Relation (13.14) really describes the solution of (13.13) (the reader is asked to verify it), but we can obtain a simpler formula.

First of all, both integrals in (13.14) can be written as

$$\iint_{|\mathbf{x}-\mathbf{x}_0|=ct} \cdots = \iint_{S^+} \cdots + \iint_{S^-} \cdots = 2\iint_{S^+} \cdots,$$

where

$$S^+ = \{(x,y,z) \in \mathbb{R}^3;\ z = \sqrt{c^2t^2 - (x-x_0)^2 - (y-y_0)^2}\},$$
$$S^- = \{(x,y,z) \in \mathbb{R}^3;\ z = -\sqrt{c^2t^2 - (x-x_0)^2 - (y-y_0)^2}\}$$

are the upper and lower hemispheres. On the upper hemisphere, we can rewrite the surface element dS as

$$dS = \sqrt{1 + \left(\frac{\partial z}{\partial x}\right)^2 + \left(\frac{\partial z}{\partial y}\right)^2}\, dx\, dy$$

$$= \sqrt{1 + \left(\frac{-(x-x_0)}{z}\right)^2 + \left(\frac{-(y-y_0)}{z}\right)^2}\, dx\, dy = \frac{ct}{z}\, dx\, dy$$

$$= \frac{ct}{\sqrt{c^2t^2 - (x-x_0)^2 - (y-y_0)^2}}\, dx\, dy.$$

Thus, formula (13.14) can be simplified to

$$u(x_0, y_0, t) = 2\frac{1}{4\pi c^2 t} \iint_D \psi(x,y) \frac{ct}{\sqrt{c^2t^2 - (x-x_0)^2 - (y-y_0)^2}}\, dx\, dy$$

$$+ 2\frac{\partial}{\partial t}\left(\frac{1}{4\pi c^2 t} \iint_D \varphi(x,y) \frac{ct}{\sqrt{c^2t^2 - (x-x_0)^2 - (y-y_0)^2}}\, dx\, dy\right),$$

where D is the disc $(x-x_0)^2 + (y-y_0)^2 \leq c^2t^2$. Thus, we can conclude that the solution of the Cauchy problem (13.13) for the wave equation on $\mathbb{R}^2$ is given by

(13.15)
$$\boxed{\begin{aligned} u(\mathbf{x}_0, t) &= \frac{1}{2\pi c} \iint_{|\mathbf{x}-\mathbf{x}_0|\leq ct} \frac{\psi(\mathbf{x})}{\sqrt{c^2t^2 - |\mathbf{x}-\mathbf{x}_0|^2}}\, d\mathbf{x} \\ &+ \frac{\partial}{\partial t}\left(\frac{1}{2\pi c} \iint_{|\mathbf{x}-\mathbf{x}_0|\leq ct} \frac{\varphi(\mathbf{x})}{\sqrt{c^2t^2 - |\mathbf{x}-\mathbf{x}_0|^2}}\, d\mathbf{x}\right). \end{aligned}}$$

Here we have $\mathbf{x} = (x,y)$ and $\mathbf{x}_0 = (x_0, y_0)$.

Let us notice the main difference between Kirchhoff's formula (13.4) for the three-dimensional problem and formula (13.15) for the two-dimensional problem. This difference concerns the domain of integration: in the former case, it is just the *spherical* surface $|x - x_0| = ct$, however, in the latter case, we integrate over the *whole disc* $|x - x_0| \leq ct$. It means that Huygens' principle does not hold true in two dimensions! For instance, in the ideal case, waves caused by a pebble thrown onto the water level propagate at a certain speed c. At the same time, every point of the water level once reached by the front wave stays in the wave motion for an infinitely long time. We could see new and new circles appearing on the water-level forever. However, the wave equation is only an approximate model and the real situation is more complicated.

Another—fictitious—example considers life in "Flatland". In such a "world" (which is only two-dimensional), any sound propagates not at the given speed c, but at all speeds less or equal to c, and thus it is heard forever. So the listener hears at one moment a mix of words the speaker has said at different times.

It can be shown that the method of spherical means can be applied in any *odd* dimension greater or equal to three, and thus Huygens' principle holds true there. Conversely, it is false in any *even* dimension.

Example 13.3 A simple example that illustrates the different wave propagation in various dimensions is the "unit hammer blow". Let us solve the problem

$$\begin{cases} u_{tt} = c^2 \Delta u, & x \in \mathbb{R}^N, \ t > 0, \\ u(x,0) = \varphi(x), & u_t(x,0) = \psi(x) \end{cases}$$

with

$$\varphi(x) \equiv 0, \quad \psi(x) = \begin{cases} 1, & |x| < a, \\ 0, & |x| > a, \end{cases}$$

choosing $N = 1, 2$ and 3. For $N = 1$ the solution is given by d'Alembert's formula, for $N = 2$ we use (13.15), and for $N = 3$ the solution is described by Kirchhoff's formula (13.4). We can observe the following behavior:

$N = 1$: At time t ($> \frac{a}{c}$), the front wave reaches the point $|x| = ct + a$. At the point $|x| = ct - a$, the wave achieves its maximal displacement (equal to $\frac{a}{c}$) and stays constant on the whole interval $|x| < ct - a$. The front wave propagates at speed c, but its influence is evident at all points $|x| < ct + a$. For details, see Example 4.5.

$N = 2$: At time t, the front wave reaches the point $|x| = ct + a$, then the wave achieves its maximum (of order $\frac{1}{\sqrt{t}}$), and, for $|x| \to 0$, it decreases as $\frac{1}{\sqrt{(ct)^2 - |x|^2}}$. The wave has a sharp front, but it has not sharp tail. As in one dimension, the nonzero initial condition at $|x| < a$ results in nonzero displacement at all points $|x| < ct + a$.

$N = 3$: At time t, again, the front wave reaches the point $|x| = ct + a$. The maximal displacement $\frac{a^2}{4c^2 t}$ is achieved at $|x| = ct$, and then the wave decreases again to zero position at $|x| = ct - a$. The whole wave propagates at speed c and does not change its shape—a nonzero initial condition at $|x| < a$ causes a nonzero displacement only at points $ct - a \leq |x| < ct - a$.

The different behavior in these three cases is sketched in Figure 13.2. □

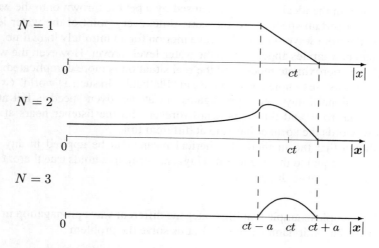

Figure 13.2 *"Hammer blow" in one, two and three dimensions.*

13.4 Wave with sources in $\mathbb{R}^3$

We will use the *operator method* for solving the non-homogeneous Cauchy problem

(13.16) $$\begin{aligned} u_{tt} - c^2 \Delta u &= f(\boldsymbol{x}, t), \quad \boldsymbol{x} = (x, y, z) \in \mathbb{R}^3, \ t > 0, \\ u(\boldsymbol{x}, 0) &= \varphi(\boldsymbol{x}), \ u_t(\boldsymbol{x}, 0) = \psi(\boldsymbol{x}). \end{aligned}$$

Let us denote by $u_H(\boldsymbol{x}, t)$ the solution of the homogeneous problem (i.e., (13.16) with $f \equiv 0$). We have found in Section 13.2 that such a solution can be written in the form

$$u_H(\boldsymbol{x}_0, t) = (\partial_t S(t)\varphi)(\boldsymbol{x}_0) + (S(t)\psi)(\boldsymbol{x}_0),$$

where S is the so called *source operator* given by the formula

(13.17) $$(S(t)\psi)(\boldsymbol{x}_0) = \frac{1}{4\pi c^2 t} \iint_{|\boldsymbol{x} - \boldsymbol{x}_0| = ct} \psi(\boldsymbol{x}) \, \mathrm{d}S.$$

The idea of the operator method is exactly the same as in Section 4.3. Again, it can be shown that the influence of the right-hand side f in problem (13.16) can be described by the term

$$u_P(\boldsymbol{x}_0, t) = \int_0^t S(t - s) f(\boldsymbol{x}_0, s) \, \mathrm{d}s.$$

Section 13.4 Wave with sources in $\mathbb{R}^3$

Hence, after substitution,

$$u_P(\boldsymbol{x}_0, t) = \int_0^t \frac{1}{4\pi c^2(t-s)} \iint_{|\boldsymbol{x}-\boldsymbol{x}_0|=c(t-s)} f(\boldsymbol{x}, s)\, dS\, ds,$$

and, using the relation $s = t - \frac{1}{c}|\boldsymbol{x} - \boldsymbol{x}_0|$ on the sphere of integration, we obtain

(13.18) $$u_P(\boldsymbol{x}_0, t) = \frac{1}{4\pi c} \int_0^t \iint_{|\boldsymbol{x}-\boldsymbol{x}_0|=c(t-s)} \frac{f(\boldsymbol{x}, t - \frac{1}{c}|\boldsymbol{x} - \boldsymbol{x}_0|)}{|\boldsymbol{x} - \boldsymbol{x}_0|} dS\, ds.$$

Here the domain of integration is, in fact, the jacket of a four-dimensional space-time cone with its vertex at $(\boldsymbol{x}_0, t)$ and the base formed by the ball $|\boldsymbol{x} - \boldsymbol{x}_0| \leq ct$. Thus, we can rewrite the expression in (13.18) into a triple integral obtaining

(13.19) $$u_P(\boldsymbol{x}_0, t) = \frac{1}{4\pi c} \iiint_{|\boldsymbol{x}-\boldsymbol{x}_0|\leq ct} \frac{f(\boldsymbol{x}, t - \frac{1}{c}|\boldsymbol{x} - \boldsymbol{x}_0|)}{|\boldsymbol{x} - \boldsymbol{x}_0|} d\boldsymbol{x}.$$

(The reader is asked to justify it.) Due to linearity of the equation, the final solution of (13.16) is the sum of u_H and u_P:

$$\boxed{\begin{aligned} u(\boldsymbol{x}_0, t) &= \frac{1}{4\pi c^2 t} \iint_{|\boldsymbol{x}-\boldsymbol{x}_0|=ct} \psi(\boldsymbol{x})\, dS + \frac{\partial}{\partial t}\left(\frac{1}{4\pi c^2 t} \iint_{|\boldsymbol{x}-\boldsymbol{x}_0|=ct} \varphi(\boldsymbol{x})\, dS \right) \\ &\quad + \frac{1}{4\pi c} \iiint_{|\boldsymbol{x}-\boldsymbol{x}_0|\leq ct} \frac{f(\boldsymbol{x}, t - \frac{1}{c}|\boldsymbol{x} - \boldsymbol{x}_0|)}{|\boldsymbol{x} - \boldsymbol{x}_0|} d\boldsymbol{x}. \end{aligned}}$$

Remark 13.4 Let us compare (13.19) with the stationary solution of the same problem, that is, the solution u_{stat} of the Poisson problem

$$-c^2 \Delta u = f$$

on the whole space $\mathbb{R}^3$. Using formula (11.18) without the boundary term and with the choice $G(\boldsymbol{x}, \boldsymbol{x}_0) = \frac{1}{4\pi c|\boldsymbol{x}-\boldsymbol{x}_0|}$, we obtain

(13.20) $$u_{\text{stat}}(\boldsymbol{x}_0) = \frac{1}{4\pi c} \iiint_{\mathbb{R}^3} \frac{f(\boldsymbol{x})}{|\boldsymbol{x} - \boldsymbol{x}_0|} d\boldsymbol{x}.$$

(The reader is asked to verify that it really solves the Poisson equation on $\mathbb{R}^3$.) As we can see, evolution formula (13.19) differs from the bounded stationary solution (13.20) just at its "retarded" time by the amount $\frac{1}{c}|\boldsymbol{x} - \boldsymbol{x}_0|$.

13.5 Characteristics, Singularities, Energy and Principle of Causality

Now we focus on the qualitative properties of the wave equation and its solution following from the equation itself but not from the formula which expresses the solution.

Characteristics. Like in one dimension, we can introduce the notion of characteristics, but now we speak about *characteristic surfaces*. The fundamental one arises if we rotate a one-dimensional characteristic line $x - x_0 = c(t - t_0)$ around the $t = t_0$ axis. We thus obtain a cone in the four-dimensional space-time (a "hypercone"):

$$(13.21) \qquad |\boldsymbol{x} - \boldsymbol{x}_0| = \sqrt{(x - x_0)^2 + (y - y_0)^2 + (z - z_0)^2} = c|t - t_0|.$$

This set is called the *characteristic cone* or the *light cone* at the point $(\boldsymbol{x}_0, t_0)$. We can imagine it as the union of all (light) rays emanating from the point $(\boldsymbol{x}_0, t_0)$ at the speed c, that is $|d\boldsymbol{x}/dt| = c$ (see Figure 13.3 for $\mathbb{R}^2$ illustration). For a fixed t, the light cone reduces to a sphere. The interior of the cone, that is $|\boldsymbol{x} - \boldsymbol{x}_0| < c|t - t_0|$, is called the *solid light cone*; it consists of the *future* and *past* half cone. Thus, for a fixed t, the future is the ball that contains all points that can be reached at time t by a particle traveling from $(\boldsymbol{x}_0, t_0)$ at a speed less than c.

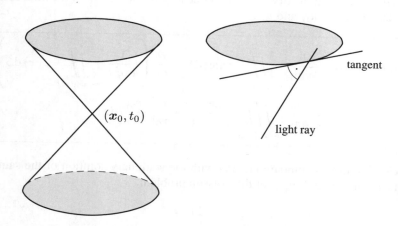

Figure 13.3 *Light cone at a point* $(\boldsymbol{x}_0, t_0)$, $\boldsymbol{x}_0 \in \mathbb{R}^2$, *and orthogonality of light rays to the sphere* $|\boldsymbol{x} - \boldsymbol{x}_0| = c|t - t_0|$.

It can be proved (see Figure 13.3 for $\mathbb{R}^2$ illustration) that the light rays of the light cone are all orthogonal to the spheres $|\boldsymbol{x} - \boldsymbol{x}_0| = c|t - t_0|$. Now we will use this property to generalize the notion of the characteristic surface. Let S be any surface in the space-time and let us denote its time slices by S_t. Notice that each slice S_t is an ordinary two-dimensional surface in $\mathbb{R}^3$. We say that S is a *characteristic surface* if it is a union of light rays which are all orthogonal to the time slices S_t.

Section 13.5 Characteristics, Singularities, Energy and Principle of Causality

Similarly to one dimension, the fundamental property of characteristic surfaces is that they are the only surfaces that can carry singularities of the solutions of the wave equation. We only recall that by a singularity we understand a point of discontinuity of the solution or of some of its derivatives.

Energy. Another property of the wave equation that remains valid in the same way as in one dimension is the conservation of energy. Indeed, if we multiply the wave equation by u_t and integrate it over $\mathbb{R}^3$, we obtain

$$(13.22) \quad 0 = \iiint_{\mathbb{R}^3} (u_{tt} - c^2 \Delta u) u_t \, d\boldsymbol{x}$$

$$= \iiint_{\mathbb{R}^3} (\tfrac{1}{2} u_t^2 + \tfrac{1}{2} c^2 |\nabla u|^2)_t \, d\boldsymbol{x} - \iiint_{\mathbb{R}^3} c^2 \nabla \cdot (u_t \nabla u) \, d\boldsymbol{x}.$$

If we rewrite the last integral as

$$\iiint_{\mathbb{R}^3} c^2 \nabla \cdot (u_t \nabla u) \, d\boldsymbol{x} = \lim_{r \to \infty} \iiint_{B_r(0)} c^2 \nabla \cdot (u_t \nabla u) \, d\boldsymbol{x},$$

where $B_r(0)$ is the ball centered at the origin with radius r, and use the Divergence Theorem, we obtain

$$\iiint_{\mathbb{R}^3} c^2 \nabla \cdot (u_t \nabla u) \, d\boldsymbol{x} = \lim_{r \to \infty} \iint_{\partial B_r(0)} c^2 u_t \nabla u \cdot \boldsymbol{n} \, d\boldsymbol{x}.$$

If we assume that the derivatives of $u(\boldsymbol{x}, t)$ tend to zero for $|\boldsymbol{x}| \to \infty$, then the last integral vanishes. Hence, (13.22) reduces to

$$0 = \iiint_{\mathbb{R}^3} (\tfrac{1}{2} u_t^2 + \tfrac{1}{2} c^2 |\nabla u|^2)_t \, d\boldsymbol{x}.$$

Moreover, if we change the order of integration and time differentiation, we obtain

$$0 = \frac{\partial}{\partial t} \iiint_{\mathbb{R}^3} (\tfrac{1}{2} u_t^2 + \tfrac{1}{2} c^2 |\nabla u|^2) \, d\boldsymbol{x}.$$

Since the term $\iiint \tfrac{1}{2} u_t^2 \, d\boldsymbol{x}$ corresponds to the *kinetic energy* E_k and the term $\iiint \tfrac{1}{2} c^2 |\nabla u|^2 \, d\boldsymbol{x}$ represents the *potential energy* E_p, we can conclude that the *total energy* $E = E_k + E_p$ is a constant function with respect to time t.

Principle of Causality. We already know (from Huygens' principle and the solution formula) that the solution of the N-dimensional Cauchy problem for the wave equation at a point $(\boldsymbol{x}_0, t_0)$ depends on the values of the initial displacement $\varphi(\boldsymbol{x})$ and the initial velocity $\psi(\boldsymbol{x})$ for $\boldsymbol{x}$ belonging to the sphere $|\boldsymbol{x} - \boldsymbol{x}_0| = ct_0$ if N is odd ($N \geq 3$), and $\boldsymbol{x}$ belonging to the whole ball $|\boldsymbol{x} - \boldsymbol{x}_0| \leq ct_0$ if N is even. However, a similar (though a little bit weaker) information follows directly from the wave equation itself.

In particular, we can formulate the so called *principle of causality*:

Theorem 13.5 *The value of $u(\mathbf{x}_0, t_0)$ can depend only on the values of $\varphi(\mathbf{x})$ and $\psi(\mathbf{x})$ for $\mathbf{x}$ from the ball $|\mathbf{x} - \mathbf{x}_0| \leq ct_0$.*

Idea of proof (cf. Strauss [19]) We use the same approach as in one dimension (see Section 10.1). We consider the three-dimensional case; however, the idea is applicable in any dimension. We take the wave equation and multiply it by u_t. After standard calculations and assuming that all derivatives make sense, we obtain

$$\begin{aligned} 0 &= u_{tt}u_t - c^2\Delta u u_t \\ &= (\frac{1}{2}u_t^2 + \frac{1}{2}c^2|\nabla u|^2)_t - c^2\nabla \cdot (u_t \nabla u) \\ &= (\frac{1}{2}u_t^2 + \frac{1}{2}c^2|\nabla u|^2)_t + (-c^2 u_t u_x)_x + (-c^2 u_t u_y)_y + (-c^2 u_t u_z)_z \\ &= \operatorname{div} \mathbf{f}, \end{aligned}$$

where $\mathbf{f}$ is a four-dimensional vector

$$\mathbf{f} = (-c^2 u_t u_x, -c^2 u_t u_y, -c^2 u_t u_z, \frac{1}{2}(u_t^2 + c^2|\nabla u|^2)).$$

Now, we integrate the equation $\operatorname{div} \mathbf{f} = 0$ over a solid cone frustum F, which is a piece of the solid light cone in the four-dimensional space-time. If we use the four-dimensional Divergence Theorem, we can write

$$\begin{aligned} 0 &= \iiiint_F \operatorname{div} \mathbf{f} = \iiint_{\partial F} \mathbf{f} \cdot \mathbf{n} \, dV \\ &= \iiint_{\partial F} [\frac{1}{2}n_4(u_t^2 + c^2|\nabla u|^2) - n_1(c^2 u_t u_x) - n_2(c^2 u_t u_y) - n_3(c^2 u_t u_z)] \, dV, \end{aligned}$$

where ∂F denotes the boundary of F and $\mathbf{n} = (n_1, n_2, n_3, n_4)$ is the unit outward normal vector to ∂F with components n_i, $i = 1, \ldots, 4$, in directions x, y, z, t. The rest of the proof is the same as in one dimension (cf. Section 10.1). Now ∂F is three-dimensional and consists of the top T, the bottom B and the jacket K (see Figure 13.4 for $\mathbb{R}^2$ illustration). Thus, the integral splits into three parts

$$\iiint_{\partial F} = \iiint_T + \iiint_B + \iiint_K = 0.$$

On the top T, the normal vector has the upward direction $\mathbf{n} = (0, 0, 0, 1)$ and the corresponding integral reduces to

$$\iiint_T (\frac{1}{2}u_t^2 + \frac{1}{2}c^2|\nabla u|^2) \, d\mathbf{x}.$$

Section 13.5 Characteristics, Singularities, Energy and Principle of Causality

Similarly, on the bottom B, the normal vector has the downward direction, that is $\mathbf{n} = (0,0,0,-1)$ and we have

$$\iiint_B -(\tfrac{1}{2}u_t^2 + \tfrac{1}{2}c^2|\nabla u|^2)\,d\boldsymbol{x} = -\iiint_B (\tfrac{1}{2}\psi^2 + \tfrac{1}{2}c^2|\nabla\varphi|^2)\,d\boldsymbol{x}.$$

On the jacket K, we cannot argue so simply, but it can be proved that the corresponding integral is positive or zero (see, e.g., Strauss [19]). Using these facts, we obtain the inequality

(13.23) $$\iiint_T (\tfrac{1}{2}u_t^2 + \tfrac{1}{2}c^2|\nabla u|^2)\,d\boldsymbol{x} \leq \iiint_B (\tfrac{1}{2}\psi^2 + \tfrac{1}{2}c^2|\nabla\varphi|^2)\,d\boldsymbol{x}.$$

Now, let us assume that the functions φ and ψ are zero on B. Inequality (13.23) implies that $\tfrac{1}{2}u_t^2 + \tfrac{1}{2}c^2|\nabla u|^2 = 0$ on T, and thus $u_t \equiv \nabla u \equiv 0$ on T. Moreover, since this result holds true for a frustum of an arbitrary height, we obtain that u_t and ∇u are zero (and thus u constant) in the entire solid cone. And since $u = 0$ on B, we can conclude $u \equiv 0$ in the entire cone. In particular, this implies that if we take two solutions u_1, u_2 with the same initial conditions on B, then $u_1 \equiv u_2$ in the entire solid cone. ∎

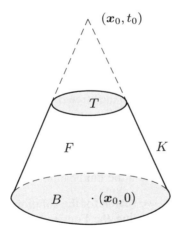

Figure 13.4 *Solid cone frustum F.*

Remark 13.6 We can state the "converse" assertion to the Principle of Causality: the initial conditions φ, ψ at the point $\boldsymbol{x}_0$ can influence the solution only in the solid light cone with its vertex at $(\boldsymbol{x}_0, 0)$. (Notice that this statement as well as the Principle of Causality hold true even for the nonhomogeneous wave equation.) We can also meet the terminology which we already know from one dimension. The past solid cone is called the *domain of dependence* and the future solid cone is called the *domain of influence* of the point (x_0, t_0).

13.6 Wave on Bounded Domains, Fourier Method

In the rest of this chapter we study initial boundary value problems for the wave equation. In general, we consider the problem

$$\begin{aligned}
&u_{tt}(\boldsymbol{x},t) = c^2 \Delta u(\boldsymbol{x},t), & & \boldsymbol{x} \in \Omega,\ t > 0, \\
&u(\boldsymbol{x},t) = h_1(\boldsymbol{x},t), & & \boldsymbol{x} \in \Gamma_1, \\
&\frac{\partial u}{\partial n}(\boldsymbol{x},t) = h_2(\boldsymbol{x},t), & & \boldsymbol{x} \in \Gamma_2, \\
&\frac{\partial u}{\partial n}(\boldsymbol{x},t) + au(\boldsymbol{x},t) = h_3(\boldsymbol{x},t), & & \boldsymbol{x} \in \Gamma_3, \\
&u(\boldsymbol{x},0) = \varphi(\boldsymbol{x}),\ u_t(\boldsymbol{x},0) = \psi(\boldsymbol{x}).
\end{aligned}$$

As usual, Ω denotes a domain in $\mathbb{R}^N$, φ, ψ, h_i, $i = 1, 2, 3$ are given functions, a is a given constant, and $\Gamma_1 \cup \Gamma_2 \cup \Gamma_3 = \partial \Omega$.

We start with an explanation of the physical meaning of the boundary conditions. If we model a vibrating membrane, $u = u(x, y, t)$ corresponds to the displacement of the membrane and the Dirichlet boundary condition on Γ_1 describes the shape of the fixed frame on which the membrane is fastened. If h_1 is not a constant, then the frame is warped. The Neumann boundary condition on Γ_2 determines the "slope" of the membrane on the boundary. In particular, the *homogeneous* Neumann boundary condition (i.e., $h_2 \equiv 0$) corresponds to the "free rim" of the membrane, which is free to flap. The Robin boundary condition on Γ_3 can describe a flexible rim of the membrane.

If we use the three-dimensional wave equation as a model of sound waves in a fluid with $u = u(x, y, z, t)$ being the fluid density, then the most common boundary condition is the homogeneous Neumann boundary condition. It corresponds to the situation when the domain has rigid walls and the fluid cannot penetrate them.

As in the previous chapters, we search for the solution of the initial boundary value problems for the wave equation using the Fourier method. Since the main idea, as well as the basic scheme, coincide completely with those for the case of the diffusion equation, we do not repeat them here in detail and refer the reader to Section 12.3. We confine ourselves only to several examples which illustrate some interesting phenomena or situations which were not treated in the previous chapter.

Example 13.7 (Rectangular Membrane.) We start with a simple situation. Let us consider a two-dimensional wave equation describing a vibrating membrane fastened on a rectangular frame which is fixed in zero position. At the beginning, let the membrane be pulled up at the center point and then released. The corresponding model can

Section 13.6 Wave on Bounded Domains, Fourier Method

have the form

(13.24) $$\begin{cases} u_{tt} = c^2(u_{xx} + u_{yy}), & (x,y) \in (0,a) \times (0,b), \ t > 0, \\ u(0,y,t) = u(a,y,t) = u(x,0,t) = u(x,b,t) = 0, \\ u(x,y,0) = \varphi(x,y), \ u_t(x,y,0) = 0, \end{cases}$$

where

(13.25) $$\varphi(x,y) = \begin{cases} xy, & 0 \leq x < a/2, \ 0 \leq y < b/2, \\ x(b-y), & 0 \leq x < a/2, \ b/2 \leq y \leq b, \\ (a-x)y, & a/2 \leq x \leq a, \ 0 \leq y < b/2, \\ (a-x)(b-y), & a/2 \leq x \leq a, \ b/2 \leq y \leq b. \end{cases}$$

The shape of the initial displacement is depicted in Figure 13.5 for the data $a = 2$, $b = 3$.

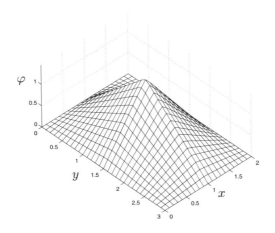

Figure 13.5 *Initial condition (13.25) with the choice $a = 2$, $b = 3$.*

First of all, we separate the time and space variables: $u(x,y,t) = V(x,y)T(t)$. Substituting into the equation in (13.24), we obtain a couple of separated equations:

(13.26) $V_{xx} + V_{yy} + \lambda V = 0$, $0 < x < a, \ 0 < y < b$,

(13.27) $T'' + \lambda c^2 T = 0$, $t > 0$,

where λ is a constant. Moreover, $V = V(x,y)$ satisfies the homogeneous boundary conditions

(13.28) $V(0,y) = V(a,y) = V(x,0) = V(x,b) = 0$, $0 < x < a, \ 0 < y < b$.

As we already know from Example 12.3, problem (13.26), (13.28) can be solved by the Fourier method and it yields the eigenvalues

$$\lambda_{mn} = \left(\frac{n\pi}{a}\right)^2 + \left(\frac{m\pi}{b}\right)^2, \quad m,n \in \mathbb{N},$$

and the corresponding orthogonal system of eigenfunctions
$$V_{mn}(x,y) = \sin\frac{n\pi x}{a}\sin\frac{m\pi y}{b}.$$
If we go back to the time equation (13.27), we obtain
$$T_{mn} = A_{mn}\cos(c\sqrt{\lambda_{mn}}t) + B_{mn}\sin(c\sqrt{\lambda_{mn}}t).$$
(Recall that all the eigenvalues λ_{mn} are positive!) Hence, the solution of the original problem can be written in the form of the double Fourier series
$$u(x,y,t) = \sum_{n=1}^{\infty}\sum_{m=1}^{\infty}\left[A_{mn}\cos(c\sqrt{\lambda_{mn}}t) + B_{mn}\sin(c\sqrt{\lambda_{mn}}t)\right]\sin\frac{n\pi x}{a}\sin\frac{m\pi y}{b}.$$
This function satisfies the required initial conditions in (13.24) provided these are also expandable into a Fourier series with respect to the system $\{V_{mn}(x,y)\}$. In our case, this means
$$\varphi(x,y) = \sum_{n=1}^{\infty}\sum_{m=1}^{\infty} A_{mn}\sin\frac{n\pi x}{a}\sin\frac{m\pi y}{b},$$
$$\psi(x,y) \equiv 0 = \sum_{n=1}^{\infty}\sum_{m=1}^{\infty} c\sqrt{\lambda_{mn}}B_{mn}\sin\frac{n\pi x}{a}\sin\frac{m\pi y}{b}.$$
The latter relation implies $B_{mn} = 0$ for all $m, n \in \mathbb{N}$. Using the orthogonality of the eigenfunctions, we can determine the coefficients A_{mn} as
$$A_{mn} = \frac{\int_0^a\int_0^b \varphi(x,y)V_{mn}(x,y)\,dy\,dx}{\int_0^a\int_0^b V_{mn}^2(x,y)\,dy\,dx}.$$
Substituting for V_{mn} and for φ from (13.25), we can calculate
$$A_{mn} = \frac{4}{ab}\int_0^a\int_0^b \varphi(x,y)\sin\frac{n\pi x}{a}\sin\frac{m\pi y}{b}\,dx\,dy$$
$$= \frac{4}{ab}\frac{4a^2b^2}{n^2m^2\pi^4}\sin\frac{n\pi}{2}\sin\frac{m\pi}{2} = \frac{16ab}{n^2m^2\pi^4}\sin\frac{n\pi}{2}\sin\frac{m\pi}{2}.$$
Now we can conclude that the solution of the initial boundary value problem (13.24) can be expressed in the form
$$u(x,y,t) = \sum_{n=1}^{\infty}\sum_{m=1}^{\infty} A_{mn}\cos(c\sqrt{\lambda_{mn}}t)\sin\frac{n\pi x}{a}\sin\frac{m\pi y}{b}.$$
Graph of the solution on several time levels is sketched in Figure 13.6. We have used the data $c = 3$, $a = 2$, $b = 3$ and the partial summation up to $n = m = 25$. The reader is invited to notice the propagation of the singularities and the reflection of the waves on the boundary.

□

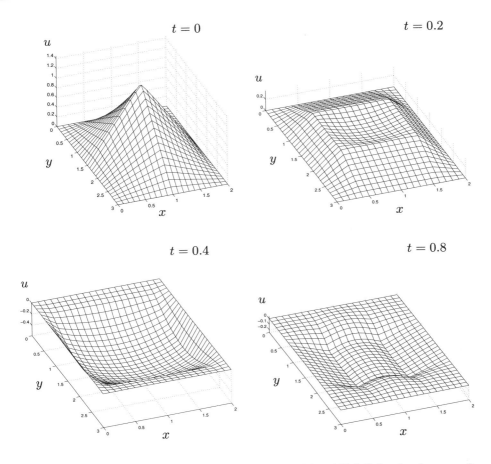

Figure 13.6 *Solution of the initial boundary value problem (13.24) for the data $c = 3$, $a = 2$, $b = 3$, on time levels $t = 0$, 0.2, 0.4, 0.8.*

The other examples deal with the wave equation on circular domains. We consider the radially symmetric as well as non-symmetric cases.

Example 13.8 (Circular Membrane—Symmetric Case.) This example is the wave analogue of Example 12.5 for the diffusion equation. This time we solve the problem

(13.29) $$\begin{cases} u_{tt} = c^2 \Delta u, & x^2 + y^2 < a^2, \; t > 0, \\ u(x,y,t) = 0, & x^2 + y^2 = a^2, \\ u(x,y,0) = \varphi(\sqrt{x^2+y^2}), & u_t(x,y,0) = \psi(\sqrt{x^2+y^2}), \end{cases}$$

which can serve as a model of a vibrating circular membrane with the frame fixed in zero position. Since the initial conditions depend only on the radius $r = \sqrt{x^2 + y^2}$, we can assume the solution to be radially symmetric, and after transformation into polar coordinates we obtain a simpler problem

(13.30)
$$\begin{cases} u_{tt} = c^2(u_{rr} + \frac{1}{r}u_r), & 0 < r < a,\ t > 0, \\ u(r,t) = 0, & r = a,\ t > 0, \\ u(r,0) = \varphi(r),\ u_t(r,0) = \psi(r), & 0 \le r < a. \end{cases}$$

If we repeat the steps of Example 12.5, we obtain the solution in the form

$$u(r,t) = \sum_{n=1}^{\infty} T_n(t) R_n(r).$$

The system of eigenfunctions R_n is given by

$$R_n(r) = J_0(\sqrt{\lambda_n} r),$$

where J_0 is the Bessel function of the first kind of order zero (see Appendix 14.2); the eigenvalues λ_n are given by

$$\lambda_n = \frac{1}{a^2} \mu_n^2, \quad n \in \mathbb{N},$$

where μ_n are the zeros of J_0. The time functions T_n are now the solutions of the equation

$$T''(t) + c^2 \lambda_n T(t) = 0.$$

Since all the eigenvalues are positive, we can write

$$T_n(t) = A_n \cos(c\sqrt{\lambda_n} t) + B_n \sin(c\sqrt{\lambda_n} t).$$

Thus, we can conclude that the solution of (13.30) assumes the form

$$u(r,t) = \sum_{n=1}^{\infty} \left[A_n \cos(c\sqrt{\lambda_n} t) + B_n \sin(c\sqrt{\lambda_n} t) \right] J_0(\sqrt{\lambda_n} r).$$

The constants A_n, B_n can be determined from the initial conditions provided these are expandable into Fourier series with respect to the system $\{R_n(r)\}$:

$$\varphi(r) = \sum_{n=1}^{\infty} A_n J_0(\sqrt{\lambda_n} r),$$

$$\psi(r) = \sum_{n=1}^{\infty} c\sqrt{\lambda_n} B_n J_0(\sqrt{\lambda_n} r).$$

Using the orthogonality of the Bessel functions (see Appendix 14.2), we obtain

$$A_n = \frac{2}{J_0'^2(\mu_n)} \int_0^1 \rho J_0(\mu_n \rho) \varphi(a\rho)\, d\rho,$$

$$B_n = \frac{2}{c\sqrt{\lambda_n} J_0'^2(\mu_n)} \int_0^1 \rho J_0(\mu_n \rho) \psi(a\rho)\, d\rho.$$

Section 13.6 Wave on Bounded Domains, Fourier Method

In particular, let us take $a = 1$, $c = 1$ and consider the initial data in the form

(13.31)
$$\varphi(r) = J_0(\mu_1 r) + J_0(\mu_2 r),$$
$$\psi(r) = 0.$$

Then, obviously, $B_n = 0$ for all $n \in \mathbb{N}$, and

$$A_1 = A_2 = 1, \quad A_n = 0, \quad n = 3, 4, \ldots$$

The corresponding solution can be then written as

$$u(x, y, t) = u(r, t) = \cos \mu_1 t \, J_0(\mu_1 r) + \cos \mu_2 t \, J_0(\mu_2 r).$$

Figure 13.7 illustrates the dependence of u on the radial coordinate r and time t. The graph of the function $u = u(x, y, t)$ on several time levels is depicted in Figure 13.8. (We recall that $x = r \cos \theta$, $y = r \sin \theta$, $\theta \in [0, 2\pi]$.) □

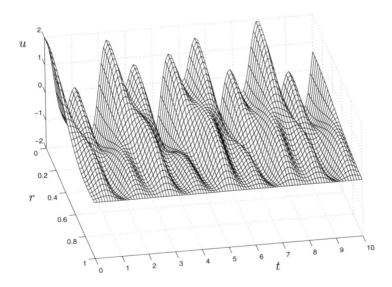

Figure 13.7 *Solution of the initial boundary value problem (13.29) with initial condition (13.31)—dependence on r and t.*

Example 13.9 (Circular Membrane—Non-Symmetric Case.) Let us consider the same problem as in the previous example, but now without any symmetry. That is, we model

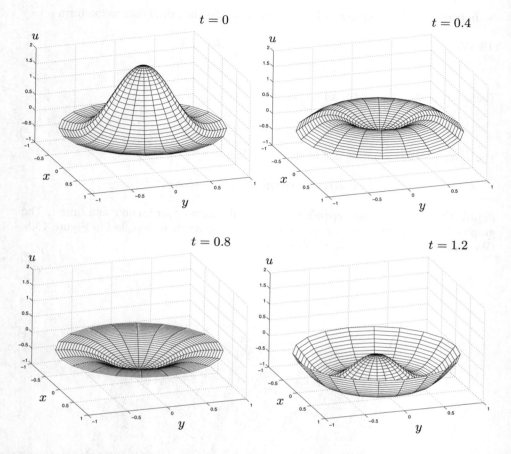

Figure 13.8 *Solution of the initial boundary value problem (13.29) with initial condition (13.31) on time levels $t = 0$, 0.4, 0.8, 1.2.*

a vibrating circular membrane with the frame fixed in zero position; the initial displacement and initial velocity are now general functions $\varphi = \varphi(x, y)$, $\psi = \psi(x, y)$:

(13.32) $\quad \begin{cases} u_{tt} = c^2 \Delta u, & x^2 + y^2 < a^2, \ t > 0, \\ u(x, y, t) = 0, & x^2 + y^2 = a^2, \\ u(x, y, 0) = \varphi(x, y), & u_t(x, y, 0) = \psi(x, y). \end{cases}$

As in the previous examples, we find the solution using the Fourier method. Since the domain is circular, we have to transform the problem again into polar coordinates, which provides the required rectangularity. However, now we have to use the general (non-symmetric) transformation formula (6.1) for the Laplace operator. Thus, (13.32) becomes

Section 13.6 Wave on Bounded Domains, Fourier Method

(13.33)
$$\begin{cases} u_{tt} = c^2(u_{rr} + \frac{1}{r}u_r + \frac{1}{r^2}u_{\theta\theta}), & 0 < r < a,\ 0 \le \theta < 2\pi,\ t > 0, \\ u(r,\theta,t) = 0, & r = a,\ 0 \le \theta < 2\pi,\ t > 0, \\ u(r,\theta,0) = \varphi(r,\theta),\ u_t(r,\theta,0) = \psi(r,\theta), & 0 \le r < a,\ 0 \le \theta < 2\pi. \end{cases}$$

In the first step, we separate the time and space variables:
$$u(r,\theta,t) = V(r,\theta)T(t),$$
and since the spatial problem will be solved again by the Fourier method, we can separate also $V(r,\theta) = R(r)\Theta(\theta)$. Thus, we have
$$u(r,\theta,t) = R(r)\Theta(\theta)T(t)$$
and the standard argument leads to
$$\frac{T''}{c^2 T} = -\lambda \quad \text{and} \quad \frac{R''}{R} + \frac{R'}{rR} + \frac{\Theta''}{r^2 \Theta} = -\lambda.$$

The separation in the latter equation results in
$$\frac{\Theta''}{\Theta} = -\nu \quad \text{and} \quad \lambda r^2 + \frac{r^2 R''}{R} + \frac{rR'}{R} = \nu.$$

Obviously, Θ must satisfy the periodic boundary conditions $\Theta(0) = \Theta(2\pi)$, $\Theta'(0) = \Theta'(2\pi)$, which implies
$$\nu_n = n^2, \quad \Theta_n(\theta) = A_n \cos n\theta + B_n \sin n\theta, \quad n = \mathbb{N} \cup \{0\}.$$

Hence, the radial equation assumes the form

(13.34)
$$r^2 R'' + rR' + (\lambda r^2 - n^2)R = 0,$$

which is the *Bessel equation of order n*. As follows from Appendix 14.2, its bounded solutions have the form
$$R(r) = J_n(\sqrt{\lambda}r),$$
where J_n is the *Bessel function of the first kind of order n*. The boundary condition in (13.33) gives $R(a) = 0$, which implies
$$\lambda_{mn} = \frac{1}{a^2}\mu_{mn}^2, \quad n \in \mathbb{N} \cup \{0\},\ m \in \mathbb{N},$$
where μ_{mn} are positive zeros of J_n. Thus, we can write $R(r) = R_{mn}(r) = J_n(\sqrt{\lambda_{mn}}r)$. Inserting $\lambda = \lambda_{mn}$ into the time equation, we obtain
$$T_{mn} = C_{mn}\cos c\sqrt{\lambda_{mn}}t + D_{mn}\sin c\sqrt{\lambda_{mn}}t.$$

Using the expressions for R, Θ, and T, we can conclude that the solution of the wave equation on the disc with homogeneous Dirichlet boundary condition has the form

$$u(r,\theta,t) = \sum_{n=0}^{\infty}\sum_{m=1}^{\infty} J_n(\sqrt{\lambda_{mn}}r)(A_{mn}\cos n\theta + B_{mn}\sin n\theta)\cos c\sqrt{\lambda_{mn}}t$$
$$+ \sum_{n=0}^{\infty}\sum_{m=1}^{\infty} J_n(\sqrt{\lambda_{mn}}r)(\overline{A}_{mn}\cos n\theta + \overline{B}_{mn}\sin n\theta)\sin c\sqrt{\lambda_{mn}}t.$$

Here we write A_{mn} instead of $A_n C_{mn}$, and similarly for B_{mn}, $\overline{A}_{mn}$, $\overline{B}_{mn}$. To determine these coefficients, we use the initial conditions. We illustrate this process on a simple example.

Let us consider problem (13.33) with the initial conditions

(13.35) $$\begin{cases} \varphi(r,\theta) = (a^2 - r^2)r\sin\theta, \\ \psi(r,\theta) = 0. \end{cases}$$

The zero initial velocity implies that all coefficients in the sine series (with respect to time variable) are zero, that is, $\overline{A}_{mn} = \overline{B}_{mn} = 0$ for all $n \in \mathbb{N}\cup\{0\}$, $m \in \mathbb{N}$. Thus the solution formula reduces to

$$u(r,\theta,t) = \sum_{n=0}^{\infty}\sum_{m=1}^{\infty} J_n(\sqrt{\lambda_{mn}}r)(A_{mn}\cos n\theta + B_{mn}\sin n\theta)\cos c\sqrt{\lambda_{mn}}t.$$

Setting $t = 0$, we obtain

(13.36) $$\varphi(r,\theta) = \sum_{n=0}^{\infty}\sum_{m=1}^{\infty} J_n(\sqrt{\lambda_{mn}}r)(A_{mn}\cos n\theta + B_{mn}\sin n\theta).$$

Notice that this is a Fourier series of the function φ with respect to the system of functions $\{J_n(\sqrt{\lambda_{mn}}r)\cos n\theta,\ J_n(\sqrt{\lambda_{mn}}r)\sin n\theta\}_{mn}$. Let us rewrite (13.36) as

$$\varphi(r,\theta) = \underbrace{\sum_{m=1}^{\infty} A_{m0} J_0(\sqrt{\lambda_{m0}}r)}_{=:A_0(r)} + \sum_{n=1}^{\infty}\underbrace{\left(\sum_{m=1}^{\infty} A_{mn} J_n(\sqrt{\lambda_{mn}}r)\right)}_{=:A_n(r)}\cos n\theta$$
$$+ \sum_{n=1}^{\infty}\underbrace{\left(\sum_{m=1}^{\infty} B_{mn} J_n(\sqrt{\lambda_{mn}}r)\right)}_{=:B_n(r)}\sin n\theta.$$

Section 13.6 Wave on Bounded Domains, Fourier Method

For a fixed r we have

$$A_0(r) = \frac{1}{2\pi} \int_0^{2\pi} \varphi(r,\theta)\,d\theta,$$

$$A_n(r) = \frac{1}{\pi} \int_0^{2\pi} \varphi(r,\theta) \cos n\theta\,d\theta,$$

$$B_n(r) = \frac{1}{\pi} \int_0^{2\pi} \varphi(r,\theta) \sin n\theta\,d\theta.$$

Substituting $\varphi(r,\theta) = (a^2 - r^2) r \sin\theta$ and using the orthogonality of trigonometric functions, we obtain $A_0 = A_n = 0$ for all $n \in \mathbb{N}$, and $B_n = 0$ for $n = 2, 3, \ldots$. The only nonzero coefficient is B_1:

$$B_1 = \sum_{m=1}^{\infty} B_{m1} J_1(\sqrt{\lambda_{m1}}\,r) = \frac{1}{\pi} \int_0^{2\pi} (a^2 - r^2) r \sin^2\theta\,d\theta.$$

Using the properties of Bessel functions (see Appendix 14.2), we obtain

(13.37) $\quad B_{m1} = \dfrac{2}{\pi a^2 J_2^2(\mu_{m1})} \displaystyle\int_0^a \int_0^{2\pi} (a^2 - r^2) r \sin^2\theta\, J_1\!\left(\mu_{m1}\dfrac{r}{a}\right) r\,d\theta\,dr$

$\qquad\qquad = \dfrac{2}{a^2 J_2^2(\mu_{m1})} \displaystyle\int_0^a (a^2 - r^2) r^2 J_1\!\left(\mu_{m1}\dfrac{r}{a}\right) dr.$

We recall that $\lambda_{m1} = (\mu_{m1}/a)^2$, and μ_{m1} are positive roots of the Bessel function J_1. Hence, we can conclude that the solution of problem (13.32) or (13.33) with the initial conditions (13.35) is given by

(13.38) $\quad\boxed{u(r,\theta,t) = \sin\theta \sum_{m=1}^{\infty} B_{m1} J_1\!\left(\mu_{m1}\dfrac{r}{a}\right) \cos \mu_{m1}\dfrac{ct}{a}}$

with B_{m1} given by (13.37).

The solution (13.38) on various time levels is plotted in Figure 13.9. Here we have put $a = 1$, $c = 1$, and used the partial sum up to $m = 3$.

$\square$

In the last example we add another spatial dimension. However, we stick to the simplest, i.e., radially symmetric, case.

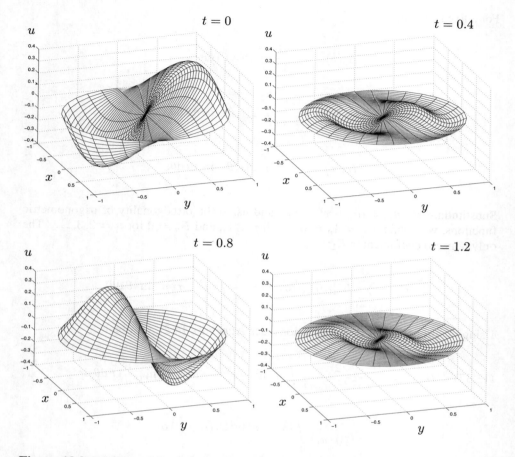

Figure 13.9 *Solution of the initial boundary value problem (13.32) with initial condition (13.35) on time levels $t = 0$, 0.4, 0.8, 1.2.*

Example 13.10 (Vibrations in Ball—Symmetric Case.) Let us consider vibrations in a ball with fixed boundary, and let the initial data depend only on the radius $r = \sqrt{x^2 + y^2 + z^2}$. That is, we solve the initial boundary value problem for the wave equation with the Dirichlet boundary condition

(13.39)
$$\begin{cases} u_{tt} = c^2 \Delta u, & x^2 + y^2 + z^2 < a^2, \ t > 0, \\ u(x, y, z, t) = 0, & x^2 + y^2 + z^2 = a^2, \\ u(x, y, z, 0) = \varphi(\sqrt{x^2 + y^2 + z^2}), \\ u_t(x, y, z, 0) = \psi(\sqrt{x^2 + y^2 + z^2}). \end{cases}$$

The geometry of the domain inspires us to transform problem (13.39) into spherical coordinates r, φ, θ. Moreover, since the data do not depend on the angles φ, θ, we

Section 13.6 Wave on Bounded Domains, Fourier Method

can expect the solution to be radially symmetric as well. Thus, the Laplace operator reduces to the simple form $\Delta u = u_{rr} + \frac{2}{r}u_r$, and (13.39) becomes a problem in two variables t and r:

(13.40)
$$\begin{cases} u_{tt} = c^2(u_{rr} + \frac{2}{r}u_r), & 0 < r < a,\ t > 0, \\ u(r,t) = 0, & r = a,\ t > 0, \\ u(r,0) = \varphi(r),\ u_t(r,0) = \psi(r), & 0 \le r < a. \end{cases}$$

To solve it, we again use the Fourier method. The separation of variables $u(r,t) = R(r)T(t)$ leads to a pair of ODEs

(13.41)
$$T'' + \lambda c^2 T = 0,$$

(13.42)
$$R'' + \frac{2}{r}R' + \lambda R = 0.$$

The radial equation can be simplified by introducing a new function $Y(r)$:

$$Y(r) = rR(r).$$

Then (13.42) becomes
$$Y''(r) + \lambda Y(r) = 0$$

and, for $\lambda > 0$, its solutions are $Y(r) = C\cos\sqrt{\lambda}r + D\sin\sqrt{\lambda}r$. Thus, we obtain

$$R(r) = \frac{1}{r}(C\cos\sqrt{\lambda}r + D\sin\sqrt{\lambda}r), \quad 0 < r < a.$$

Further, R must satisfy the boundary conditions

$$R(0) \text{ bounded}, \quad R(a) = 0.$$

Hence $C = 0$, since $\frac{1}{r}\cos\sqrt{\lambda}r$ is unbounded in the neighborhood of $r = 0$. (Remember that $\frac{1}{r}\sin\sqrt{\lambda}r$ is bounded and tends to $\sqrt{\lambda}$ for $r \to 0$.) The latter boundary condition implies

$$D\sin\sqrt{\lambda}a = 0,$$

which gives the eigenvalues

$$\lambda_n = \left(\frac{n\pi}{a}\right)^2, \quad n \in \mathbb{N}$$

and the corresponding system of eigenfunctions

$$R_n(r) = \frac{1}{r}\sin\frac{n\pi r}{a}, \quad n \in \mathbb{N}.$$

The time problem (13.41) has solutions $T_n(t) = A_n\cos c\sqrt{\lambda_n}t + B_n\sin c\sqrt{\lambda_n}t$, and so the radially symmetric solution of the wave equation satisfying the Dirichlet boundary condition can be written as

(13.43)
$$u(r,t) = \sum_{n=1}^{\infty}(A_n\cos\frac{n\pi ct}{a} + B_n\sin\frac{n\pi ct}{a})\frac{1}{r}\sin\frac{n\pi r}{a}$$

for $r > 0$. For the evaluation of $u(0, t)$ we use the fact that $\lim_{r \to 0} \frac{1}{r} \sin \frac{n\pi r}{a} = \frac{n\pi}{a}$ and set

$$u(0, t) = \sum_{n=1}^{\infty} \frac{n\pi}{a}\left(A_n \cos \frac{n\pi ct}{a} + B_n \sin \frac{n\pi ct}{a}\right).$$

To satisfy also the initial conditions, we have to ensure

$$\varphi(r) = \sum_{n=1}^{\infty} A_n \frac{1}{r} \sin \frac{n\pi r}{a},$$

$$\psi(r) = \sum_{n=1}^{\infty} \frac{n\pi c}{a} B_n \frac{1}{r} \sin \frac{n\pi r}{a}$$

for $r > 0$, which is equivalent to

$$r\varphi(r) = \sum_{n=1}^{\infty} A_n \sin \frac{n\pi r}{a},$$

$$r\psi(r) = \sum_{n=1}^{\infty} \frac{n\pi c}{a} B_n \sin \frac{n\pi r}{a}$$

for $r \geq 0$. Using the standard argument, the coefficients A_n, B_n can be written as

$$A_n = \frac{2}{a} \int_0^a r\varphi(r) \sin \frac{n\pi r}{a}\, dr,$$

$$B_n = \frac{2}{n\pi c} \int_0^a r\psi(r) \sin \frac{n\pi r}{a}\, dr.$$

Since we cannot easily plot u as a function of all variables x, y, z, t, Figure 13.10 depicts only the values of the solution (13.43) in dependence on the radial coordinate r and on time t. We have chosen parameters $a = 1$, $c = 1$, zero initial velocity $\psi \equiv 0$, and the initial displacement given by

(13.44) $$\varphi(r) = J_0(\mu_1 r) + J_0(\mu_2 r).$$

(Here J_0 is the Bessel function of the first kind of order zero, see Appendix 14.2.) We have used the partial sum up to $n = 20$. The initial condition (13.44) is sketched in Figure 13.11.

□

Remark 13.11 In the previous example, we have chosen the initial displacement (13.44) since the same condition was used in Example 13.8 for the problem of a vibrating membrane. Let us have a detailed look at Figure 13.10 and compare it with Figure 13.7. The

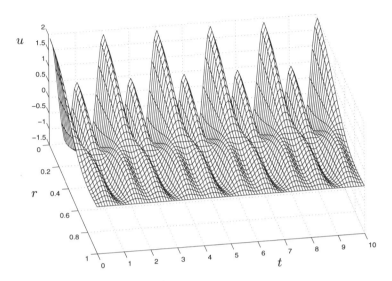

Figure 13.10 *Solution of the initial boundary value problem (13.41) with initial condition (13.44).*

former illustrates radially symmetric vibrations in a unit ball (e.g., sound waves), while the latter depicts radially symmetric vibrations in a unit disc. In both cases, we have used the same data, but the behavior is very different (notice, e.g., the shape of the propagating waves and the time period).

Another example of different behavior in two and three dimensions is illustrated in Figure 13.12. There we have depicted the solution of the wave equation in two and three dimensions. This time we have chosen (in both cases) parameters $a = 10$, $c = 1.5$, zero initial displacement $\varphi \equiv 0$, and the initial velocity given by

$$\psi(r) = 1 \quad \text{for } 0 \leq r \leq 1, \quad \psi(r) = 0 \quad \text{elsewhere}.$$

(We treated the wave equation with the same initial conditions—but on the whole plane and space—in Example 13.3.) In two dimensions, this corresponds to the situation when we hit the membrane by a unit circular hammer. We can see that the signal propagates along the characteristics, and when it reaches any point, the displacement there never vanishes. On the other hand, in three dimensions, the signal comes and fades away. This corresponds to the fact that, in two dimensions, the (non-reflected) solution at a point (x_0, t_0) is influenced by the initial values from the *whole disc*

$$(x - x_0)^2 + (y - y_0)^2 \leq (ct_0)^2,$$

while, in three dimensions, only the initial values from the *spherical surface*

$$(x - x_0)^2 + (y - y_0)^2 + (z - z_0)^2 = (ct_0)^2$$

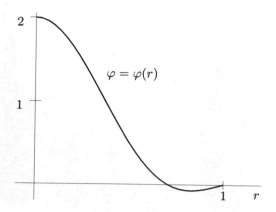

Figure 13.11 *The initial displacement (13.44).*

are relevant. Notice also the reflection on the boundary $r = a$ and the effect of the Principle of Causality, which is valid in any dimension.

Remark 13.12 The radial equation (13.42) is a special case (with $n = 0$) of the general equation

$$(13.45) \qquad r^2 R'' + 2rR' + (\lambda r^2 - n(n+1))R = 0,$$

which appears in non-symmetric problems in a ball. It can be shown that solutions of (13.45) have the form

$$R(r) = \sqrt{\frac{\pi}{2\sqrt{\lambda} r}} J_{n+\frac{1}{2}}(\sqrt{\lambda} r),$$

where $J_{n+\frac{1}{2}}$ is the Bessel function of the first kind of (non-integer) order $n + \frac{1}{2}$, see Appendix 14.2. Setting $n = 0$, the solution of (13.42) can be written as

$$R(r) = \sqrt{\frac{\pi}{2\sqrt{\lambda} r}} J_{\frac{1}{2}}(\sqrt{\lambda} r).$$

Using expression (14.6) for Bessel functions, we obtain

$$\begin{aligned}
J_{\frac{1}{2}}(x) &= \sum_{k=0}^{\infty} (-1)^k \frac{(x/2)^{\frac{1}{2}+2k}}{k! \Gamma(\frac{1}{2}+k+1)} \\
&= \sqrt{\frac{2}{\pi x}} \sum_{k=0}^{\infty} \frac{(-1)^k}{(2k+1)!} x^{2k+1} = \sqrt{\frac{2}{\pi x}} \sin x.
\end{aligned}$$

Section 13.6   Wave on Bounded Domains, Fourier Method

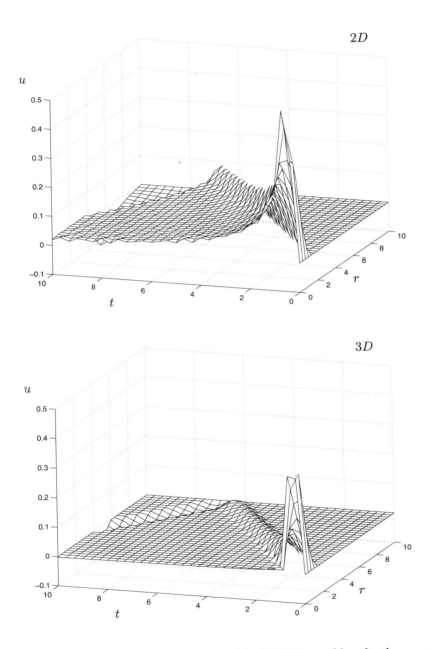

Figure 13.12 *Radially symmetric solutions of the Dirichlet problem for the wave equation in 2D and 3D with the initial condition $\psi(r) = 1$ for $0 \leq r \leq 1$ and zero otherwise.*

Thus, the solution of (13.42) assumes the simple form

$$R(r) = \frac{1}{\sqrt{\lambda}r} \sin \sqrt{\lambda}r,$$

as we have already derived in a different way.

13.7 Exercises

1. Find all three-dimensional plane waves; that is, all solutions of the wave equation in the form $u(\mathbf{x}, t) = f(\mathbf{k} \cdot \mathbf{x} - ct)$, where $\mathbf{k}$ is a fixed vector and f is a function of one variable.

    [either $|\mathbf{k}| = 1$ or $u(\mathbf{x}, t) = a + b(\mathbf{k} \cdot \mathbf{x} - ct)$, where a, b are arbitrary constants]

2. Verify that $(c^2 t^2 - x^2 - y^2 - z^2)^{-1}$ satisfies the wave equation except on the light cone.

3. Prove that $\Delta(\bar{u}) = (\overline{\Delta u}) = \bar{u}_{rr} + \frac{2}{r}\bar{u}_r$ for any function $u = u(x, y, z)$. Here $r = \sqrt{x^2 + y^2 + z^2}$ is the spherical coordinate.

    [Hint: Write Δu in spherical coordinates and show that the angular terms have zero average on spheres centered at the origin.]

4. Using Kirchhoff's formula, solve the wave equation in three dimensions with the initial data $\varphi(x, y, z) \equiv 0$, $\psi(x, y, z) = y$.

    $[u(x, y, z, t) = ty]$

5. Solve the wave equation in three dimensions with the initial data $\varphi(x, y, z) \equiv 0$, $\psi(x, y, z) = x^2 + y^2 + z^2$. Search for a radially symmetric solution and use the substitution $v(r, t) = ru(r, t)$.

6. Solve the wave equation in three dimensions with initial conditions $\varphi(\mathbf{x}) \equiv 0$, $\psi(\mathbf{x}) = A$ for $|\mathbf{x}| < \rho$ and $\psi(\mathbf{x}) = 0$ for $|\mathbf{x}| > \rho$, where A is a constant. This problem is an analogue of the hammer blow solved in Section 4.2.

    $[u(\mathbf{x}, t) = \frac{A}{4cr}(\rho - (r - ct)^2)$ for $|\rho - ct| \leq r \leq \rho + ct$,
    $u(\mathbf{x}, t) = At$ for $r \leq \rho - ct$, and $u(\mathbf{x}, t) = 0$ elsewhere]

7. Solve the wave equation in three dimensions with initial conditions $\varphi(\mathbf{x}) = A$ for $|\mathbf{x}| < \rho$, $\varphi(\mathbf{x}) = 0$ for $|\mathbf{x}| > \rho$ and $\psi(\mathbf{x}) \equiv 0$, where A is a constant. Where has the solution jump discontinuities?

    [Hint: Differentiate the solution from Exercise 6.]

    $[u(\mathbf{x}, t) = A$ for $r < \rho - ct$, $u(\mathbf{x}, t) = A(r - ct)/2r$ for $|\rho - ct| < r < \rho + ct$, and $u(\mathbf{x}, t) = 0$ for $r > \rho + ct]$

8. Use Kirchhoff's formula and the reflection method to solve the wave equation in the half-space $\{(x, y, z, t); z > 0\}$ with the Neumann condition $\partial u/\partial z = 0$ on $z = 0$, and with initial conditions $\varphi(x, y, z) \equiv 0$ and arbitrary $\psi(x, y, z)$.

9. Why doesn't the method of spherical means work for two-dimensional waves?

10. Suppose that we do not know d'Alembert's formula and solve the one-dimensional wave equation with the initial data $\varphi(x) \equiv 0$ and arbitrary $\psi(x)$ using the descent method. That is, think of $u(x, t)$ as a solution of the two-dimensional equation independent of the y variable.

11. Consider the wave equation with the boundary condition $\partial u/\partial n + b\, \partial u/\partial t = 0$, $b > 0$, and show that its energy decreases.

12. Consider the equation $u_{tt} - c^2 \Delta u + m^2 u = 0$, $m > 0$, known as the *Klein-Gordon equation*. Show that its energy is constant.

13. Prove the uniqueness of the classical solution of the wave equation on $\mathbb{R}^3$. Use the conservation of energy applied to the difference of two solutions.

14. Find the value $u(0, 0, 0, t)$ of the solution of the wave equation

$$u_{tt} - \Delta u = g$$

in three spatial variables if

(a) $\varphi(x, y, z) = f(x^2 + y^2 + z^2)$, $\psi \equiv 0$, $g \equiv 0$,
(b) $\varphi \equiv 0$, $\psi(x, y, z) = f(x^2 + y^2 + z^2)$, $g \equiv 0$,
(c) $\varphi \equiv 0$, $\psi \equiv 0$, $g(x, y, z, t) = f(x^2 + y^2 + z^2)$.

[If we denote $v(t) = u(0, 0, 0, t)$, then a) $v(t) = f(c^2 t^2) + 2c^2 t^2 f'(c^2 t^2)$; b) $v(t) = tf(c^2 t^2)$; c) $v(t) = \int_0^t (t - \tau) f(c^2(t - \tau)^2) d\tau$.]

15. Consider the equation

$$u_{tt} = c^2 u_{xx} - b u_t + a u_{yy}$$

on the rectangle $(0, a) \times (0, b)$ with boundary conditions

$$u(0, y, t) = 0, \quad u_x(a, y, t) = 0,$$
$$u_y(x, 0, t) = 0, \quad u(x, b, t) = 0.$$

Find the corresponding separated ODEs and boundary conditions.

16. Separate the PDE

$$u_{tt} = c^2(u_{xx} + u_{yy} + u_{zz}) - (u_x + u_y)$$

into the corresponding ODEs.

17. Solve the two-dimensional wave equation on the unit square with the coefficient $c = \frac{1}{\pi}$, homogeneous Dirichlet boundary conditions, and the following initial conditions:

   (a) $\varphi(x,y) = \sin 3\pi x \, \sin \pi y$, $\psi(x,y) = 0$,

   $$[u(x,y,t) = \sin 3\pi x \, \sin \pi y \, \cos \sqrt{10}\, t]$$

   (b) $\varphi(x,y) = \sin \pi x \, \sin \pi y$, $\psi(x,y) = \sin \pi x$,

   (c) $\varphi(x,y) = x(1-x)y(1-y)$, $\psi(x,y) = 2 \sin \pi x \, \sin 2\pi y$,

   $$[u(x,y,t) = \sum_{l=0}^{\infty} \sum_{k=0}^{\infty} \frac{64 \cos \sqrt{(2k+1)^2 + (2l+1)^2}\, t}{\pi^6 (2k+1)^3 (2l+1)^3} \sin(2k+1)\pi x \, \sin(2l+1)\pi y$$
   $$+ \frac{2}{\sqrt{5}} \sin \pi x \, \sin 2\pi y \, \sin \sqrt{5}\, t]$$

   (d) $\varphi(x,y) = x(1 - e^{x-1})y(1 - y^2)$, $\psi(x,y) = 0$.

18. Solve the two-dimensional wave equation on a disc of radius a with homogeneous Dirichlet boundary condition. Use the following data:

   (a) $a = 2$, $c = 1$, $\varphi(r) = 0$, $\psi(r) = 1$,

   $$[u(r,t) = 4 \sum_{n=1}^{\infty} \frac{J_0(\mu_n r/2)}{\mu_n^2 J_1(\mu_n)} \sin \frac{\mu_n}{2} t]$$

   (b) $a = 1$, $c = 10$, $\varphi(r) = 1 - r^2$, $\psi(r) = 1$,

   (c) $a = 1$, $c = 1$, $\varphi(r) = 0$, $\psi(r) = J_0(\mu_3 r)$,

   (d) $a = 1$, $c = 1$, $\varphi(r) = J_0(\mu_3 r)$, $\psi(r) = 1 - r^2$.

   $$[u(r,t) = J_0(\mu_3 r) \cos \mu_3 t + 8 \sum_{n=1}^{\infty} \frac{J_0(\mu_n r)}{\mu_n^4 J_1(\mu_n)} \sin \mu_n t]$$

19. Solve the two-dimensional wave equation on a disc of radius a with homogeneous Dirichlet boundary condition. Use the following data:

   (a) $a = 1$, $c = 1$, $\varphi(r,\theta) = (1 - r^2) r^2 \sin 2\theta$, $\psi(r,\theta) = 0$,

   $$[u(r,\theta,t) = 24 \sum_{n=1}^{\infty} \frac{J_2(\mu_{n2} r)}{\mu_{n2}^3 J_3(\mu_{n2})} \sin 2\theta \, \cos \mu_{n2} t]$$

   (b) $a = 1$, $c = 1$, $\varphi(r,\theta) = 0$, $\psi(r,\theta) = (1 - r^2) r^2 \sin 2\theta$,

   $$[u(r,\theta,t) = 24 \sum_{n=1}^{\infty} \frac{J_2(\mu_{n2} r)}{\mu_{n2}^4 J_3(\mu_{n2})} \sin 2\theta \, \sin \mu_{n2} t]$$

   (c) $a = 1$, $c = 1$, $\varphi(r,\theta) = 1 - r^2$, $\psi(r,\theta) = J_0(r)$.

20. Consider a thin rectangular plate of length a and width b and describe its vibrations for the following data: $a = \frac{\pi}{2}$, $b = \pi$, $c = 1$, boundary conditions

   $$u(0, y, t) = 0, \quad u_x(a, y, t) = 0,$$
   $$u_y(x, 0, t) = 0, \quad u(x, b, t) = 0,$$

and initial conditions

$$u(x,y,0) = \begin{cases} y\sin x, & 0 \leq x \leq a,\ 0 \leq y < \frac{b}{2}, \\ (y-b)\sin x, & 0 \leq x \leq a,\ \frac{b}{2} \leq y \leq b, \end{cases}$$

and

$$u_t(x,y,0) = \begin{cases} x(\cos y + 1), & 0 \leq x < \frac{a}{2},\ 0 \leq y \leq b, \\ (x-a)(\cos y + 1), & \frac{a}{2} \leq x \leq a,\ 0 \leq y \leq b. \end{cases}$$

21. Consider a thin vibrating rectangular membrane of length $\frac{3\pi}{2}$ and width $\frac{\pi}{2}$. Suppose the sides $x = 0$ and $y = 0$ are fixed and the other two sides are free. Given zero initial velocity and the initial displacement $\varphi(x,y) = (\sin x)(\sin y)$, determine the time-dependent solution and plot its graph on several time levels.

22. Solve the problem of a vibrating circular membrane of radius 1 with fixed boundary, zero initial velocity, and the initial displacement described by $f(r)\sin 2\theta$.

23. Solve the problem of a vibrating circular membrane of radius π with free boundary, zero initial velocity, and the initial displacement described by $f(r)\cos\theta$.

24. Consider vertical vibrations of a circular sector $0 < \theta < \frac{\pi}{4}$ with radius 2. Determine the solution if the boundary conditions are

   (a) $u(r,0,t) = 0$, $u(r,\frac{\pi}{4},t) = 0$, $u(2,\theta,t) = 0$.
   (b) $u_\theta(r,0,t) = 0$, $u_\theta(r,\frac{\pi}{4},t) = 0$, $u(2,\theta,t) = 0$.

   In both cases, assume zero initial velocity and the initial displacement as a function of the radius and the angle.

25. Solve the problem

$$\begin{cases} u_{tt} - c^2(u_{xx} + u_{yy}) = f(x,y)\sin\omega t, & (x,y) \in \Omega = (0,a) \times (0,b),\ t > 0, \\ u(x,y,t) = 0, & (x,y) \in \partial\Omega, \\ u(x,y,0) = 0,\ u_t(x,y,0) = 0. \end{cases}$$

Consider separately the nonresonance case $\omega \neq \omega_{mn} = c\sqrt{(\frac{m\pi}{a})^2 + (\frac{n\pi}{b})^2}$ for all $m, n \in \mathbb{N}$, and the resonance case $\omega = \omega_{m_0 n_0}$ for some (m_0, n_0).

26. Find all solutions of the wave equation of the form $u = e^{i\omega t}f(r)$ that are finite at the origin. Here $r = \sqrt{x^2 + y^2}$.

$$[u(r,t) = Ae^{-i\omega t}J_0(\tfrac{\omega r}{c})]$$

Exercises 1–12, 26 are taken from Strauss [19], 14 from Barták et al. [4], 15, 16, 20–24 from Keane [12], 17–19 from Asmar [3], and 25 from Stavroulakis and Tersian [18].

14 Appendix

14.1 Sturm-Liouville problem

When dealing with the Fourier method we have met parametric boundary value problems for the second order ODEs, whose solutions (usually, sines and cosines) form a complete orthogonal system. This fact plays the crucial role in finding the solution of the original PDE problem in the form of an infinite series. These properties are not typical only for sines and cosines, but also for more general functions which arise as solutions of the so called *Sturm-Liouville boundary value problem*

(14.1)
$$\begin{array}{l} -(p(x)y')' + q(x)y = \lambda r(x)y, \quad a < x < b, \\ \alpha_0 y(a) + \beta_0 y'(a) = 0, \\ \alpha_1 y(b) + \beta_1 y'(b) = 0. \end{array}$$

Here, $\alpha_0^2 + \beta_0^2 > 0$, $\alpha_1^2 + \beta_1^2 > 0$ (i.e., at least one number of each pair is nonzero), and λ is a parameter.

We say that (14.1) forms a *regular Sturm-Liouville problem*, if $[a, b]$ is a closed finite interval and the following *regularity conditions* are fulfilled: $p(x)$, $p'(x)$, $q(x)$ and $r(x)$ are continuous real functions on $[a, b]$, and $p(x) > 0$, $r(x) > 0$ for $a \leq x \leq b$.

Any value of the parameter $\lambda \in \mathbb{R}$ for which the nontrivial solution of problem (14.1) exists is called an *eigenvalue*. The corresponding nontrivial solution is called an *eigenfunction* related to the eigenvalue λ.

Now, we summarize the main important properties of the eigenvalues and eigenfunctions of regular Sturm-Liouville problems:

- The eigenvalues of problem (14.1) are all real, and form an increasing infinite sequence
$$\lambda_1 < \lambda_2 < \lambda_3 < \cdots < \lambda_n < \cdots \to \infty.$$

- To each eigenvalue λ_n there corresponds a unique (up to a nonzero multiple) eigenfunction $y_n(x)$, which has exactly $n-1$ zeros in (a, b). (Notice that any multiple of an eigenfunction is also an eigenfunction.) Moreover, between two consecutive zeros of $y_n(x)$ there is exactly one zero of $y_{n+1}(x)$.

- If $y_n(x)$ and $y_m(x)$ are two eigenfunctions corresponding to two different eigenvalues λ_n and λ_m, then
$$\int_a^b r(x) y_n(x) y_m(x) \, \mathrm{d}x = 0$$

(i.e., y_n and y_m are linearly independent and orthogonal with respect to the weight function $r(x)$).

- Any piecewise smooth function defined on (a, b) is expandable into Fourier series with respect to the eigenfunctions y_n, that is

$$f(x) = \sum_{n=1}^{\infty} F_n y_n(x),$$

where F_n are the Fourier coefficients defined by the relation

$$F_n = \frac{\int_0^l r(x) f(x) y_n(x)\,dx}{\int_0^l r(x) y_n^2(x)\,dx}.$$

Moreover, the series converges at $x \in [a, b]$ to $f(x)$ if $f(x)$ is continuous at x, and it converges to $\frac{1}{2}(f(x^+) + f(x^-))$ if $f(x)$ has a jump discontinuity at x (here $f(x^+)$ and $f(x^-)$ are one sided limits at x).

We usually say that the eigenfunctions $y_n(x)$ form a complete orthogonal set.

In some cases and under additional conditions, the above mentioned properties are valid also for the so called *singular Sturm-Liouville problems*, see, e.g., [20]. In our text we deal with only one such case: the parametric Bessel equation (see Examples 12.5, 13.8, 13.9 and Remark 14.1 below).

14.2 Bessel Functions

In Chapters 12 and 13 we have met a special case of the so called *Bessel equation of order n*

(14.2) $$x^2 y'' + x y' + (x^2 - n^2) y = 0$$

or

$$y'' + \frac{1}{x} y' + \left(1 - \frac{n^2}{x^2}\right) y = 0, \quad x \neq 0.$$

Here n is a nonnegative constant (not necessarily integer; but for our purposes, we usually consider $n \in \mathbb{N}$). Equation (14.2) is a linear second-order ODE and thus it must have a pair of linearly independent solutions, which can be searched in the form

$$y(x) = \sum_{k=0}^{\infty} a_k x^{k+\alpha}, \quad a_0 \neq 0.$$

Below, we find concrete values of a_k and α. Substituting back into (14.2), we obtain

$$x^2 \sum_{k=0}^{\infty} (k+\alpha)(k+\alpha-1) a_k x^{k+\alpha-2}$$

$$+ x \sum_{k=0}^{\infty} (k+\alpha) a_k x^{k+\alpha-1} + (x^2 - n^2) \sum_{k=0}^{\infty} a_k x^{k+\alpha} = 0$$

or (after a simplification and canceling the term x^α)

$$\sum_{k=0}^{\infty}(k+\alpha-n)(k+\alpha+n)a_k x^k + \sum_{k=0}^{\infty} a_k x^{k+2} = 0.$$

The second sum can be rewritten as

$$\sum_{k=0}^{\infty} a_k x^{k+2} = \sum_{k=2}^{\infty} a_{k-2} x^k$$

and we obtain the equation

(14.3) $$\sum_{k=0}^{\infty}(k+\alpha-n)(k+\alpha+n)a_k x^k + \sum_{k=2}^{\infty} a_{k-2} x^k = 0.$$

Thus, the coefficients at the particular powers of x must satisfy

$$k=0: \quad (\alpha-n)(\alpha+n)a_0 = 0,$$
$$k=1: \quad (1+\alpha-n)(1+\alpha+n)a_1 = 0,$$
$$k\geq 2: \quad (k+\alpha-n)(k+\alpha+n)a_k + a_{k-2} = 0.$$

Since we require $a_0 \neq 0$, the first equation implies

$$\alpha = n \quad \text{or} \quad \alpha = -n.$$

The second equation must hold for any nonnegative n, and thus $a_1 = 0$. The third equation leads to the *recursive formula*

(14.4) $$a_k = \frac{-1}{(k+\alpha-n)(k+\alpha+n)} a_{k-2}.$$

Since $a_1 = 0$, it follows that

$$a_3 = a_5 = \cdots = a_{2k+1} = \cdots = 0,$$

and the only nonzero coefficients can be written as

(14.5) $$a_{2k} = \frac{-1}{(2k+\alpha-n)(2k+\alpha+n)} a_{2k-2}$$

$$= \frac{(-1)^k a_0}{2^{2k} k!(1+n)(2+n)(3+n)\ldots(k+n)}.$$

Thus, making the conventional choice $a_0 = 2^{-n}/n!$ and taking $\alpha = n$, we obtain the first solution of the Bessel equation

(14.6) $$J_n(x) = \sum_{k=0}^{\infty} (-1)^k \frac{(x/2)^{n+2k}}{k!(n+k)!},$$

Section 14.2 Bessel Functions

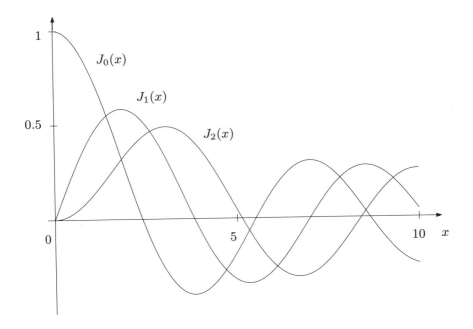

Figure 14.1 *Bessel functions of the first kind for $n = 0, 1, 2$.*

which is called the *Bessel function of the first kind of order n*. (If $n \notin \mathbb{N}$, we have to replace the factorial $(n + k)!$ by the so called Gamma function $\Gamma(n + k + 1)$, see, e.g., Abramowitz, Stegun [1].) Several Bessel functions of the first kind are sketched in Figure 14.1. Notice that all these functions are finite even at the singular point $x = 0$!

It can be shown that the second linearly independent solution of the Bessel equation has the form

$$Y_n(x) = \lim_{q \to n} \frac{J_q(x) \cos q\pi - J_{-q}(x)}{\sin q\pi}$$

and is known as the *Bessel function of the second kind of order n*. This function is unbounded at $x = 0$. In fact, it behaves like x^{-n} near $x = 0$ for $n > 0$. In the case $n = 0$, it looks like logarithm near the origin. Several Bessel functions of the second kind are sketched in Figure 14.2.

Without proofs, we state here the basic properties of Bessel functions which we have used in this book.

- As we can see in Figures 14.1, 14.2, each Bessel function has a countable number of distinct positive roots μ_k, $k = 1, 2, \ldots$.

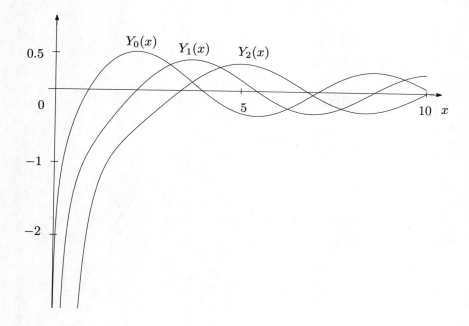

Figure 14.2 *Bessel functions of the second kind for* $n = 0, 1, 2$.

- For any $n \geq 0$, the system of functions $\{\sqrt{x} J_n(\mu_{nk} x)\}_{k=1}^{\infty}$ is orthogonal on $[0, 1]$:

$$\int_0^1 x J_n(\mu_{nk} x) J_n(\mu_{nj} x) \mathrm{d}x = 0 \quad \text{for } j \neq k$$

and

$$\int_0^1 x J_n^2(\mu_{nk} x) \mathrm{d}x = \frac{1}{2} J_{n+1}^2(\mu_{nk}).$$

Here μ_{nk}, $k = 1, 2, \ldots$, are again the positive roots of $J_n(x)$.

- For any $n \geq 0$,

$$\frac{\mathrm{d}}{\mathrm{d}x}[x^n J_n(x)] = x^n J_{n-1}(x) \quad \text{and} \quad \frac{\mathrm{d}}{\mathrm{d}x}[x^{-n} J_n(x)] = -x^{-n} J_{n+1}(x).$$

In particular, we have $\frac{\mathrm{d}}{\mathrm{d}x} J_0(x) = -J_1(x)$.

Remark 14.1 In Examples 12.5, 13.8 and 13.9 we have met the so called *parametric form of the Bessel equation*

(14.7) $$x^2 y'' + x y' + (\lambda x^2 - n^2) y = 0$$

Section 14.2 Bessel Functions

with the boundary conditions

(14.8) $\qquad\qquad\qquad y(0)$ finite, $\quad y(a) = 0.$

If μ_{nk} are positive roots of $J_n(x)$, then problem (14.7), (14.8) is solvable for the values $\lambda = \lambda_{nk} = \frac{\mu_{nk}^2}{a^2}$, and the corresponding solutions are $J_n(\frac{\mu_{nk}}{a} x)$.

Notice that equation (14.7) can be written as

$$-(xy')' + \frac{n^2}{x} y = \lambda x y,$$

which is (together with the boundary conditions) nothing else but the Sturm-Liouville problem on $(0, a)$ with $p(x) = r(x) = x$ and $q(x) = \frac{n^2}{x}$. Since $q(x)$ is not defined and $p(x)$ vanishes at $x = 0$, this problem is a singular one. The Sturm-Liouville theory then implies some of the above mentioned properties of Bessel functions, namely, that the functions $J_n(\frac{\mu_{nk}}{a} x)$ form a complete orthogonal system on $(0, a)$ with respect to the weight function $r(x) = x$.

More properties of Bessel functions can be found, e.g., in Abramowitz, Stegun [1].

Some Typical Problems Considered in This Book

Transport equation (Chapter 3)
$$u_t + cu_x = 0$$

Transport equation with decay (Chapter 3)
$$u_t + cu_x + \lambda u = f$$

Wave equation

- Cauchy problem on $\mathbb{R}$ (Chapters 4, 9)
$$\begin{aligned} u_{tt} &= c^2 u_{xx} + f, \quad x \in \mathbb{R},\ t > 0, \\ u(x,0) &= \varphi(x),\ u_t(x,0) = \psi(x) \end{aligned}$$

- initial boundary value problem in $\mathbb{R}$ (Chapters 7, 9)
$$\begin{aligned} u_{tt} &= c^2 u_{xx} + f, \quad x \in (0,l),\ t > 0, \\ u(0,t) &= u(l,t) = 0, \\ u(x,0) &= \varphi(x),\ u_t(x,0) = \psi(x) \end{aligned}$$

- in higher dimension (Chapter 13)
$$\begin{aligned} u_{tt} &= c^2 \Delta u + f \quad \text{in } \mathbb{R}^2 \text{ or } \mathbb{R}^3,\ t > 0, \\ u(\boldsymbol{x},0) &= \varphi(\boldsymbol{x}),\ u_t(\boldsymbol{x},0) = \psi(\boldsymbol{x}) \end{aligned}$$

- initial boundary value problem in higher dimension (Chapter 13)
$$\begin{aligned} u_{tt} &= c^2 \Delta u + f \quad \text{in } \Omega,\ t > 0, \\ u(\boldsymbol{x},t) &= 0 \quad \text{on } \partial\Omega, \\ u(\boldsymbol{x},0) &= \varphi(\boldsymbol{x}),\ u_t(\boldsymbol{x},0) = \psi(\boldsymbol{x}) \end{aligned}$$

Diffusion equation

- Cauchy problem on $\mathbb{R}$ (Chapters 5, 9)
$$\begin{aligned} u_t &= k u_{xx} + f, \quad x \in \mathbb{R},\ t > 0, \\ u(x,0) &= \varphi(x) \end{aligned}$$

- initial boundary value problem in $\mathbb{R}$ (Chapters 7, 9)

$$\begin{aligned} &u_t = ku_{xx} + f, \quad x \in (0,l),\ t > 0, \\ &u(0,t) = u(l,t) = 0, \\ &u(x,0) = \varphi(x) \end{aligned}$$

- in higher dimension (Chapter 12)

$$\begin{aligned} &u_t = k\Delta u + f \quad \text{in } \mathbb{R}^2 \text{ or } \mathbb{R}^3,\ t > 0, \\ &u(\boldsymbol{x},0) = \varphi(\boldsymbol{x}) \end{aligned}$$

- initial boundary value problem in higher dimension (Chapter 12)

$$\begin{aligned} &u_t = k\Delta u + f \quad \text{in } \Omega,\ t > 0, \\ &u(\boldsymbol{x},t) = 0 \quad \text{on } \partial\Omega, \\ &u(\boldsymbol{x},0) = \varphi(\boldsymbol{x}) \end{aligned}$$

Laplace (Poisson) equation

- in $\mathbb{R}^2$ or $\mathbb{R}^3$ (Chapters 6, 11)

$$\Delta u = f \quad \text{in } \mathbb{R}^2 \text{ or } \mathbb{R}^3$$

- boundary value problems (Chapters 8, 11)

$$\begin{aligned} &\Delta u = f \quad \text{in } \Omega, \\ &u = 0 \quad \text{on } \partial\Omega \end{aligned}$$

Notation

ODE	ordinary differential equation	
PDE	partial differential equation	
$\mathbb{R}, \mathbb{R}^N$	the set of real numbers, N-dimensional Euclidean space	p. 1
$\partial \Omega$	boundary of the set Ω	p. 2
$a \cdot b$	scalar product of vectors a and b	p. 2
$n(x)$	outer normal vector at the point x	p. 2
$\operatorname{div} \phi(x) \left(= \nabla \cdot \phi = \frac{\partial \phi_1}{\partial x_1} + \cdots + \frac{\partial \phi_N}{\partial x_N}\right)$	divergence of the vector function ϕ	p. 3
$u_t \left(= \frac{\partial u}{\partial t}\right), \phi_x \left(= \frac{\partial \phi}{\partial x}\right), u_{tt} \left(= \frac{\partial^2 u}{\partial t^2}\right), \phi_{xx} \left(= \frac{\partial^2 \phi}{\partial x^2}\right), \ldots$ partial derivatives		p. 5
$\operatorname{grad} u(x) = \nabla u \left(= \left(\frac{\partial u}{\partial x_1}, \ldots, \frac{\partial u}{\partial x_N}\right)\right)$	gradient of the scalar function u	p. 6, 22
$\Delta u(x) \left(= \operatorname{div}(\operatorname{grad} u) = \frac{\partial^2 u}{\partial x_1^2} + \cdots + \frac{\partial^2 u}{\partial x_N^2}\right)$	Laplace operator	p. 6
i	imaginary unit	p. 10
$\frac{\partial u}{\partial n} (= \operatorname{grad} u \cdot n)$	derivative with respect to outer normal	p. 12
$\partial_x = \frac{\partial}{\partial x}, \partial_x^2 = \frac{\partial^2}{\partial x^2}, \ldots$	partial derivatives	p. 15
$\mathbb{C}$	the set of complex numbers	p. 16
C^1 (C^2, C^3)	the space of once (twice, three times) continuously differentiable functions	p. 42
$\operatorname{erf}(z)$	error function	p. 60
$\varphi(x-) = \lim_{t \to x-} \varphi(t), \varphi(x+) = \lim_{t \to x+} \varphi(t)$		p. 63
C^∞	the set of all functions whose partial derivatives of any order are also continuous	p. 64
$\operatorname{rot} E \left(= \nabla \times E = \left(\frac{\partial E_3}{\partial y} - \frac{\partial E_2}{\partial z}, \frac{\partial E_1}{\partial z} - \frac{\partial E_3}{\partial x}, \frac{\partial E_2}{\partial x} - \frac{\partial E_1}{\partial y}\right)\right)$ rotation of the vector field E		p. 71

$\mathbb{N}$	the set of positive integers	p. 87
$\mathcal{L}$	Laplace transform	p. 123
$\mathcal{L}^{-1}$	inverse Laplace transform	p. 123
$\mathcal{F}$	Fourier transform	p. 129
$\mathcal{F}^{-1}$	inverse Fourier transform	p. 130
$u * v$	convolution of functions u and v	p. 125, 131
$\mathcal{S}$	Schwartz set	str. 129
$\mathfrak{R}$	space-time cylinder	p. 145
$L^2(M)$	the space of all functions the second powers of which are integrable on the set M	p. 148
$f(t) = O(t^n)$	the ratio $\frac{f(t)}{t^n}$ is bounded as $t \to 0$	p. 151
$\boldsymbol{B}^t$	transposed matrix to the matrix $\boldsymbol{B}$	p. 157
meas $\partial B(\boldsymbol{0}, a)$	measure (surface) of the ball $B(\boldsymbol{0}, a)$	p. 162
$\|\nabla u\|^2 = \|\operatorname{grad} u\|^2 = u_x^2 + u_y^2 + u_z^2$		p. 162
$\boldsymbol{a} \times \boldsymbol{b}$	vector product of vectors $\boldsymbol{a}$ and $\boldsymbol{b}$	p. 176
$\Delta^2 u = \Delta(\Delta u)$	biharmonic operator	p. 176

We keep the same notation for a function u when applying the transformation of its independent variables, i.e., $u = u(x, y)$ and $u = u(r, \theta)$ for transformation from Cartesian into polar coordinates etc.

Bibliography

[1] M. ABRAMOWITZ, I.A. STEGUN, *Handbook of Mathematical Functions*, Dover Publications, Inc., New York, 1965.

[2] V. I. ARNOLD, *Lectures on Partial Differential Equations*, Springer-Verlag, Berlin Heidelberg, 2004.

[3] N. ASMAR, *Partial Differential Equations and Boundary Value Problems*, Prentice Hall, Upper Saddle River, 2000.

[4] J. BARTÁK, L. HERRMANN, V. LOVICAR, O. VEJVODA, *Partial Differential Equations of Evolution*, Ellis Horwood-SNTL, New York-Praha, 1991.

[5] P. BASSANINI, A. R. ELCRAT, *Theory and Applications of Partial Differential Equations*, Plenum Press, New York, 1997.

[6] D. BLEECKER, G. CSORDAS, *Basic Partial Differential Equations*, International Press, Cambridge, 1996.

[7] L. C. EVANS, *Partial Differential Equations*, American Mathematical Society, Providence, Rhode Island, 1998.

[8] S. J. FARLOW, *Partial Differential Equations for Scientists and Engineers*, John Wiley & Sons, Inc., New York, 1982.

[9] J. FRANCŮ, *Partial Differential Equations*, PC-DIR Real, Brno, 1998, (in Czech).

[10] F. G. FRIEDLANDER, M. JOSHI, *Introduction to the Theory of Distributions*, Cambridge University Press, Cambridge, 1999.

[11] A. FRIEDMAN, *Partial Differential Equations of Parabolic Type*, Prentice Hall, Englewood-Cliffs, 1964.

[12] M. K. KEANE, *A Very Applied First Course in Partial Differential Equations*, Prentice Hall, Upper Saddle River, 2002.

[13] J. D. LOGAN, *Applied Mathematics: A Contemporary Approach*, J. Wiley & Sons, Inc., New York, 1987.

[14] J. D. LOGAN, *Applied Partial Differential Equations*, Springer-Verlag, New York, 1998.

[15] S. MÍKA, A. KUFNER, *Partial Differential Equations I, Stationary Equations*, SNTL, Praha, 1983 (in Czech).

[16] M. H. PROTTER, H. F. WEINBERGER, *Maximum Principles in Differential Equations*, Prentice Hall, Englewood Cliffs, 1967.

[17] A. D. SNIDER, *Partial Differential Equations: Sources and Solutions*, Prentice Hall, Upper Saddle River, 1999.

[18] I. P. STAVROULAKIS, S. A. TERSIAN, *Partial Differential Equations, An Introduction with Mathematica and Maple*, World Scientific, Singapore, 1999.

[19] W. A. STRAUSS, *Partial Differential Equations: An Introduction*, John Wiley & Sons, Inc., New York, 1992.

[20] W. WALTER, *Ordinary Differential Equations*, Graduate Texts in Mathematics, Vol. 182, Springer-Verlag, New York, 1992.

[21] H. F. WEINBERGER, *A First Course in Partial Differential Equations*, Blaisdell, Waltham, Mass., 1965.

[22] E. ZAUDERER, *Partial Differential Equations of Applied Mathematics*, John Wiley & Sons, Inc., New York, 1989.

Index

balance domain, 2
balance law, 3
Bessel equation, 189, 229
Bessel function, 189, 231
boundary condition, 9
    Dirichlet, **12**, 85, 89
    homogeneous, 13
    mixed, 13
    Neumann, **12**, 91
    Newton, **12**
    nonhomogeneous, 13
    periodic, 116
    Robin, **12**, 93
boundary value problem, **12**, 113
    Dirichlet
        homogeneous, 12
    Sturm-Liouville, 98, 228
Brownian motion, 144
Burgers equation, 8

Cauchy problem, 37
    for diffusion equation, 58
    for wave equation, 41
Cauchy-Riemann conditions, 72
chaos, 144
characteristic cone, 204
characteristic coordinates, 27
characteristic equation, 116
characteristic lines, 23
characteristic surface, 204
characteristic triangle, **49**, 139
characteristics, **23**, 204
    of wave equation, 41
circular membrane, 211
classical solution, 13
coefficient
    diffusion, 6
Cole-Hopf transform, 8
compatibility condition, 13
condition
    boundary, 9
        Dirichlet, 12
        homogeneous, 13
        mixed, 13
        Neumann, 12
        Newton, 12
        nonhomogeneous, 13
        periodic, 116

        Robin, 12
    Cauchy-Riemann, 72
    compatibility, 13
    initial, 9, 13
cone
    characteristic, 204
    light, 204
    solid, 204
conservation
    of energy, 205
    of mass, 7
    of momentum, 7
conservation law
    energy, 141
    evolution, 3
    heat, 7
    mass, 38
    stationary, 4
constant
    diffusion, 57
constitutive law, 6
convection, 4, **6**, 21
convective diffusion, 104
coordinate method, 25
coordinates
    characteristic, 27

d'Alembert's formula, 42, 45
damped string, 153
damped wave equation, 53
damping
    external, 39
density, 6
    mass, 2, 37
    source, 2
differential operator
    linear, 9
diffusion, 6
    convective, 104
    on half-line, 78
diffusion coefficient, 6
diffusion constant, 57
diffusion equation, **6**, 10, 57, 178
    with sources, 65
diffusion kernel, 61, 180
diffusion process, 6
Dirac distribution, 61
Dirichlet boundary condition, **12**, 85, 89

Dirichlet principle, 162
distribution
    Dirac, 61
divergence theorem, 3, **159**, 205
domain
    balance, 2
        space-time, 2
    of dependence, **49**, 139
    of influence, **48**, 139
Duhamel's principle, 52

eigenfunction, 98, 228
eigenvalue, 98, 228
electric charge, 72
electric potential, 72
electrostatic field, 72
electrostatic potential, 159
electrostatics, **71**, 159
elliptic type, 14
energy
    heat
        internal, 7
    internal, 57
        specific, 6
    kinetic, **141**, 163, 205
    potential, **142**, 163, 205
    total, 205
energy conservation law, 141
energy method, **147**, 163
equation
    Bessel, 189, 229
    Burgers, 8
    characteristic, 116
    diffusion, **6**, 10, 57, 178
        with sources, 65
    evolution, 9
    Fisher's, 8
    heat, **7**, 57
    homogeneous, 9
    kinetic, 38
    Korteweg-deVries, 138
    Laplace, **7**, 10, 71, 113, 157
    linear, 9
    Maxwell, 71
    nonhomogeneous, 10
    nonlinear, 9
    of disperse wave, 11
    of vibrating beam, 10
    partial differential, 1
    Poisson, **7**, 10, 71, 113, 157, 168

Schrödinger, 10
    stationary, 9
    telegraph, **8**, 56
    transport, **6**, 10, 21
    wave, **7**, 37, 195
        damped, 53
        with interaction, 10
        with sources, 49
        with constant coefficients, 22
equilibrium state, 71
error function, 60
evolution conservation law, 3
evolution equation, 9
evolution process, 3
external damping, 39
external force, 39

Fick's law, **6**, 57
Fisher's equation, 8
flow
    heat, **7**, 57
    steady, 72
flow function, 1
flow parameter, 6
flow quantity, 1
flux
    heat, 6, **7**
focusing effect, 199
force
    external, 39
forced vibrations, 126
formula
    d'Alembert's, 42, 45
    integral, 117
    Kirchhoff's, 196
    Poisson, **117**, 173
    representation, 164
Fourier method, **84**, 98, 113, 183
Fourier series, 117
Fourier transform, **123**, 128
    inverse, 130
Fourier's law, **7**, 57
function
    Bessel, 189, 231
    error, 60
    flow, 1
    Green's, **61**
    harmonic, 71
        radially symmetric, 75
    holomorphic, 72

source, 3, 61
state, 1
fundamental solution, **61**, 180

Gauss complex plane, 72
Gaussian, 61
Gaussian transform, 130
general solution, 11
Green's first identity, 160, **160**
Green's function, **61**
Green's second identity, 164
Green's theorem, **50**, 140, 195
ground state, 163

harmonic function, 71
　　radially symmetric, 75
heat conductivity, **7**, 58, 179
heat conservation law, 7
heat energy
　　internal, 7
heat equation, **7**, 57
heat flow, **7**, 57
　　on half-line, 78
heat flux, 6, **7**
heat kernel, 61
heat transfer
　　lateral, 103
holomorphic function, 72
homogeneous boundary condition, 13
homogeneous equation, 9
Huygens' principle, 199
hyperbolic type, 14

identity
　　Green's first, 160, **160**
　　Green's second, 164
ill-posed problem, **14**, 144
initial boundary value problem, **13**, 78
initial condition, 9, **13**
initial displacement, **13**, 41
initial value problem, 41
initial velocity, **13**, 41
inner tension, 37
integral formula, 117
integral transform, 123
internal energy, 57
internal heat energy, 7
inverse Fourier transform, 130
inverse Laplace transform, 123
irreversible process, 144

isotropic, 73

Jacobi matrix, 73

kinetic energy, **141**, 163, 205
kinetic equation, 38
Kirchhoff's formula, 196
Korteweg-deVries equation, 138

Laplace equation, **7**, 10, 71, 113, 157
Laplace transform, 123
　　inverse, 123
Laplacian, 73
lateral heat transfer, 103
law
　　balance, 3
　　conservation
　　　　energy, 141
　　　　evolution, 3
　　　　heat, 7
　　　　mass, 38
　　　　stationary, 4
　　constitutive, 6
　　Fick's, **6**, 57
　　Fourier's, **7**, 57
　　logistic, **8**, 106
　　Newton's of motion, **7**, 38
left traveling wave, 21
light cone, 204
　　solid, 204
linear differential operator, 9
linear equation, 9
logistic law, **8**, 106

mass conservation law, 38
mass density, 2, 37
material relations, 6
mathematical model, 1
mathematical modeling, 1
maximum principle, **145**, 148
　　strong, **150**, 162
Maxwell equations, 71
mean value property, **150**, 161
membrane
　　circular, 211
　　rectangular, 208
method
　　coordinate, 25
　　energy, **147**, 163
　　Fourier, **84**, 98, 113, 183

of characteristic coordinates, 27
of characteristics, **22**, 41
of integral transforms, 123
operator, **50**, 51, 202
reflection, **78**, 84, 168, 171
mixed boundary condition, 13
model
    convection, **6**, 21
    traffic, 8

Neumann boundary condition, **12**, 91
Neumann problem, 164
Newton boundary condition, **12**
Newton's law of motion, **7**, 38
nonhomogeneous boundary condition, 13
nonhomogeneous equation, 10
nonlinear equation, 9
nonlinear transport, 21

operator
    differential
        linear, 9
    factorable, 53
    reducible, 53
    source, **52**, 66, 202
operator method, **50**, 51, 202
order of equation, 9
overdetermined problem, 14

parabolic type, 14
parameter
    flow, 6
    state, 6
partial differential equation, 1
periodic boundary condition, 116
Poisson equation, **7**, 10, 71, 113, 157, 168
Poisson formula, **117**, 173
potential
    electric, 72
    electrostatic, 159
    velocity, 72
potential energy, **142**, 163, 205
principle
    Dirichlet, 162
    Duhamel's, 52
    Huygens', 199
    maximum, **145**, 148
        strong, **150**, 162
    of causality, 48, **139**, 205

reciprocity, 167
problem
    boundary value, **12**, 113
        Dirichlet homogeneous, 12
        Sturm-Liouville, 98, 228
    Cauchy, 37
    ill-posed, **14**, 144
    initial boundary value, **13**, 78
    initial value, 41
    Neumann, 164
    overdetermined, 14
    underdetermined, 14
    unstable, 14
    well-posed, 13
process
    diffusion, 6
    evolution, 3
    irreversible, 144
    stationary, 7
propagator, 61
property
    mean value, **150**, 161

quantity
    flow, 1
    state, 1

radial symmetry, **73**, 75, 158
reciprocity principle, 167
rectangular membrane, 208
reflection method, **78**, 84, 168, 171
relations
    material, 6
representation formula, 164
right traveling wave, 21
Robin boundary condition, **12**, 93

Schrödinger equation, 10
Schwartz space, 129
series
    Fourier, 117
singular point, 43
singularity, 43
sink, 3
solid light cone, 204
solution, 11
    classical, 13
    fundamental, **61**, 180
    general, 11
source, 3

Index

source density, 2
source function, **3**, 61
source operator, **52**, 66, 202
space
    Schwartz, 129
space-time balance domain, 2
space-time cylinder, 145
specific heat capacity, 57
specific internal energy, 6
speed
    of wave propagation, **7**, 39
spherical means, 197
stability
    uniform, 147
standing waves, 43
state
    equilibrium, 71
    ground, 163
    stationary, 4
    steady, 71
state function, 1
state parameter, 6
state quantity, 1
stationary conservation law, 4
stationary equation, 9
stationary process, 7
stationary state, 4
steady flow, 72
steady state, 71
stiffness, 39
string, 37
string vibration, 37
strong maximum principle, **150**, 162
Sturm-Liouville boundary value problem, 98, 228
surface
    characteristic, 204

telegraph equation, **8**, 56
temperature
    thermodynamic, 57
theorem
    divergence, 3, **159**, 205
    Green's, **50**, 140, 195
    Weierstrass, 150
thermal conductivity, **7**, 58, 179
thermal diffusivity, **7**, 58, 179
thermodynamic temperature, 57
time interval, 2
tone
    fundamental, 88
    higher, 88
traffic model, 8
transform
    Cole-Hopf, 8
    Fourier, **123**, 128
        inverse, 130
    Gaussian, 130
    integral, 123
    Laplace, 123
        inverse, 123
transport, 6
    nonlinear, 21
    with decay, 21
    with diffusion, 58
transport equation, **6**, 10, 21
type
    elliptic, 14
    hyperbolic, 14
    parabolic, 14

underdetermined problem, 14
uniform stability, 147
uniqueness of solution, **146**, 149
unstable problem, 14

velocity potential, 72

wave
    left traveling, 21
    on half-line, 80
    right traveling, 21
wave equation, **7**, 37, 195
    damped, 53
    with interaction, 10
    with sources, 49
wave motion, 7
well-posed problem, 13